Technological Innovation Across Nations

Marina van Geenhuizen • Chihiro Watanabe
Vinnie Jauhari • Enno Masurel
Editors

Technological Innovation Across Nations

Applied Studies of Coevolutionary Development

Springer

Prof. Dr. Marina van Geenhuizen
Delft University of Technology
Faculty of Technology, Policy
and Management, and OTB Research Institute
P.O. Box 5015
2600 GA Delft
The Netherlands
m.s.vangeenhuizen@tudelft.nl

Prof. Dr. Chihiro Watanabe
Tokyo Institute of Technology
Department of Industrial Engineering
and Management
2-12-1 W9-49 Ookayama, Meguro-ku
Tokyo 152-8552
Japan
watanabe.c.aa@m.titech.ac.jp

Prof. Dr. Vinnie Jauhari
HP Labs Open Innovation Office
Global Business Park,
Tower D, 6th Floor
Gurgaon 122 002 Haryana
India
vinnie.jauhari@hp.com

Prof. Dr. Enno Masurel
Free University Amsterdam
Center of Entrepreneurship
De Boelelaan 1085
1081 HV Amsterdam
The Netherlands
emasurel@feweb.vu.nl

IIASA

ISBN 978-3-642-10118-2 e-ISBN 978-3-642-00158-1

DOI 10.1007/978-3-642-00158-1

Springer is a part of Springer Science + Business Media

Published by Springer-Verlag Berlin Heidelberg 2009

Cover design: WMXDesign GmbH, Heidelberg, Germany

Printed on acid-free paper

springer.com

Preface

Innovation is the driving force behind economic growth. The knowledge that fosters innovation builds on both systematic research and serendipity. The availability of new knowledge—be it technological or organizational (social)—is, however, insufficient. An institutional structure must be in place that enhances mutual learning among the actors involved and that stimulates cooperation, as well as competition. This recognition has inspired this volume, which focuses on national innovation systems and sectoral (technology) innovation systems that differ widely between countries, due to their diverse historical paths and distinctive cultural and societal features. With regard to theory, most chapters incorporate a mix of endogenous growth theory and evolutionary thinking. Endogenous growth theory places an emphasis on the interplay between technological knowledge and various structural characteristics of the economy and society, as well as on the results of economic growth. Concepts related to technology development, accumulation of knowledge on technology, and technology diffusion feature prominently here. In evolutionary thinking, and in particular coevolutionary thinking, the focus is on close links between technology and institutions, whereby the functionality of technology is influenced by certain dynamic interactions with institutional systems. Information technology turns out to be a clear example of such a development. Although evidence of specific coevolutionary developments are presented in our book, with regard to information and communication technology it turned out to be quite difficult to compare countries and highly diverse innovation systems due to size differences in economy and distinctive orientations within their innovation systems. Nonetheless, there are promising new alleys of research ahead, not only in conceptual terms (coevolutionary development under divergent conditions), but also in terms of refining and testing suitable indicators in the statistical analysis of the field.

This book brings together some of the papers presented at the IIASA-Tokyotech meetings that took place between 2002 and 2006 at IIASA (International Institute for Applied Systems Analysis) in Laxenburg (Austria). These meetings were chaired by Prof. Dr. Chihiro Watanabe in an efficient and particularly amiable yet scientific atmosphere. The meetings' stimulating ambience, together with the interesting results on learning, knowledge networking, technological innovation, institutional systems, and coevolutionary development convinced me to take the lead in putting together this book with the objective of distributing the results to a larger audience. I thank Dr. Henri Groot (Free University in Amsterdam, Netherlands), who assisted in making a first selection of papers, as well as my co-editors who did part of the editorial work. I am also grateful to Niki Rodousakis who has assisted in improving the language of non-native English speakers and in "streamlining" the chapters.

Delft, Delft University of Technology,
Faculty of Technology,
Policy and Management
September 2008

Prof. Dr. Marina van Geenhuizen,
Head of the Editorial Team

Contents

List of Contributors

Marina van GEENHUIZEN is a Professor at the Faculty of Technology, Policy and Management (Section Economics of Innovation) and OTB Research Institute, Delft University of Technology, The Netherlands. Her research interests include the theory and practice of regional- and urban-based new high-technology business (particularly life sciences and ICT), knowledge valorization and transfer, and national and regional technology policies, with a special focus on institutional conditions for innovation diffusion and use, and evolutionary developments. Dr. van Geenhuizen received her Ph.D. from Erasmus University (Rotterdam) in 1993 and joined Delft University of Technology in 1995. She has published around 60 articles in peer-reviewed journals and is the primary editor of six international books.

Charla GRIFFY-BROWN is an Associate Professor of Information Systems and holds the Denny Endowed Chair at Pepperdine University's Graziadio School of Business and Management. A former researcher at the Foundation for Advanced Studies on International Development, Tokyo, Dr. Griffy-Brown has also served as an Associate Professor at the Tokyo Institute of Technology. She graduated from Harvard University, is a former Fulbright Scholar, and holds a Ph.D. in Technology Management from Griffith University in Queensland, Australia. She has worked for NASA at the Kennedy Space Center and served as a consultant for the United Nation's Global Environmental Facility and the European Commission.

Vinnie JAUHARI is the Regional Leader for HP Labs Open Innovation Office in India. She was a Professor and the Head of The School of Management and Entrepreneurship at IIMT (in collaboration with Oxford Brookes University, UK). She is the Founding Editor of the Journal of Services Research and has authored six books including *From Chaos to Serenity – The Divine Perspective on Management*; *Women, Technology and Entrepreneurship*; and *Services Management*. Her research interests include management of technology and strategic management. She has completed her Master's degree in Physics and an MBA (Gold Medalist). She received her Ph.D. in Management from IIT, Delhi, and held a postdoctoral position in Sustainable Technologies at the United Nations University, Tokyo.

Jianbin JIN is a Professor and Assistant Dean at the School of Journalism and Communication, Tsinghua University. His research interests lie in empirical studies on the adoption, use, and impact of new media. He received his Ph.D. from Hong Kong Baptist University in 2002. He was a visiting scholar at the Simon Fraser University in British Columbia, Canada, and has published two books and over 30 articles and papers on Internet Studies, Media, Society and Culture, and Research Methods.

Masanobu KITA is a Professor at the Graduate School of Management, Kitakyusyu Municipal University, Japan. He received his Ph.D. from the Tokyo Institute of Technology in 2005.

Shanyu LEI is an Assistant Professor at the School of Management, Dalian University of Technology, China. He received his Ph.D. in Industrial Engineering and Management from the Tokyo Institute of Technology in 2007.

Enno MASUREL is a Professor of Sustainable Entrepreneurship and the Director of the VU Center for Entrepreneurship at the VU University, Amsterdam, The Netherlands. His main research interests are entrepreneurship and small- and medium-sized enterprises (SMEs), with a special focus on innovation. He has participated in several international seminars and published numerous international papers. In a 2006 study on publications in the field of entrepreneurship, Professor Masurel was ranked among the top three in The Netherlands and the top eight worldwide.

Danny SOETANTO is a Ph.D. candidate at the Department of Economic Innovation, Faculty of Technology, Policy and Management, Delft University of Technology, The Netherlands. He completed his M.B.A. and M.A. in Systems Engineering at the Petra Christian University, Indonesia. His research interests are in Technological Entrepreneurship. He has published numerous articles and book chapters, and has received a number of awards and honors.

Yuji TOU is an Assistant Professor at the Department of Industrial Engineering and Management, Tokyo Institute of Technology, Japan. He graduated from Tokyo Institute of Technology in 2005 with a Ph.D. in Industrial Engineering and Management.

Chihiro WATANABE is a Professor at the Department of Industrial Engineering and Management, Tokyo Institute of Technology, Japan. He is also a Senior Advisor to the Director of Technology at the International Institute for Applied Systems Analysis (IIASA) in Vienna, Austria. He graduated from Tokyo University with a Bachelor's degree in Engineering (Urban Planning) in 1968, and received his Ph.D. (Arts and Science) in 1992 from Tokyo University. Professor Watanabe's research fields include Techno-Economics, Technology Innovation and Dissemination, and Technology Policy. He has published widely in the field of Industrial Technology, Energy, and the Environment.

Frans WERKHOVEN is a Professor at the Windesheim University of Applied Sciences and is the Head of the Economic and Social Institute (ESI). He graduated from the Rijksuniversiteit Groningen and is involved in the program development of Kennispoort Regio Zwolle.

Akihisa YAMADA is an Assistant Manager at Sigmaxyz Inc., Japan. He received his Ph.D. in Industrial Engineering and Management from the Tokyo Institute of Technology in 2007.

Bing ZHU is an Associate Professor at Tsinghua University, China, and a Guest Research Scholar at the International Institute for Applied Systems Analysis (IIASA), Austria. He received his B.E. (1990) and M.E. (1997) degrees in Chemical Engineering from Tsinghua University and completed his Ph.D. (2000) in Industrial Engineering and Management at the Tokyo Institute of Technology, Japan. His research interests include Technology Innovation, Industrial Ecology, and Energy Economics.

Introduction

Marina van Geenhuizen, Chihiro Watanabe, Vinnie Jauhari, Enno Masurel

1 Introduction

Origin of the Book

This book draws on papers presented at the biannual Technical Meetings between IIASA and TiTech on the "Elucidation of the Role of Institutional Systems in Characterizing Technology Development Trajectories". Eight meetings, chaired by Professor Dr. Chihiro Watanabe (Tokyo Institute of Technology and IIASA), took place between 2002 and 2006. These meetings offered an interesting blend of applied state-of-the-art studies on the coevolutionary development of institutions, innovation diffusion, and innovative entrepreneurship, which are now being published for the first time.

Marina van Geenhuizen
Faculty of Technology Policy and Management, Delft University of Technology, PO Box 5015, 2600 GA Delft, The Netherlands, e-mail: m.s.vangeenhuizen@tudelft.nl

Chihiro Watanabe
Department of Industrial Engineering & Management, Tokyo Institute of Technology, 2-12-1, W9-49, Ookayama, Meguro-ku, Tokyo 152-8552 Japan, e-mail: chihiro@me.titech.ac.jp

Vinnie Jauhari
Region Lead, HP Labs Open Innovation Office, Tower D, 8th Floor, Mehrauli-Gurgaon Road, Gurgaon 122002, India, e-mail: vinnie.jauhari@hp.com

Enno Masurel
VU Centre for Entrepreneurship, VU University Amsterdam The Netherlands, e-mail: emasurel@feweb.vu.nl

M. V. Van Geenhuizen et al., (eds). *Technological Innovation Across Nations*,
DOI: 10.1007/978-3-642-00158-1_Intro, © Springer-Verlag Berlin Heidelberg 2009

Theoretical Concepts and Approaches

This book, by and large, can be classified as mainstream *endogenous growth theory*. Endogenous growth theory assumes growth of economies on the basis of scale and accumulation, particularly through positive externalities and spillover effects from the development of high-value added production, thus enabling companies to gain competitive advantage in a globalized economy. Much attention in endogenous growth theory is paid to the role of technology (Romer, 1990; Aghion and Howitt, 1998). Accordingly, enterprises develop knowledge and innovations through investments in research and development. These investments are seen as a production factor rather than an exogenous variable. Investments in research and development can also generate positive impacts on other companies (spillover effect). Other mainstream authors have placed strong emphasis on the relation between technology and trade (Grossman and Helpman, 1991). In endogenous growth theory, economic growth is perceived as being engaged in a two-way interaction between technology and economics (Aghion and Howitt, 1998). Endogenous growth theory rests on the interplay between technological knowledge and various structural characteristics of the economy and society, as well as on the results of economic growth. Furthermore, the theory provides deeper insights into how organizations, institutions, market structure, trade, government policy, and regulation affect and are affected by long-term growth through their impacts on economic agents' incentives to engage in new entrepreneurial ventures and innovative activity (Audretsch et al., 2006; Cohen et al., 2002).

A second and related theory this book contributes to is *evolutionary theory*. Evolutionary theory focuses on change and adaptation and on how and why economic structures change continuously (Hodgson, 1999). Key concepts in this field of study include variation and selection environments, the latter determining which variation will persist. Some studies focus on cognitive aspects of economic agents, such as routines, beliefs, objectives, imperfect trial and error, etc., which are influenced by previous experiences and the agent's environment. In this vein, selection processes take place within a heterogeneous population of firms from a given sector (Nelson and Winter, 1982; Dosi et al., 1995). Industrial sectors may also be described as systems in which agents are related and interact in a number of ways. Selection processes among industrial sectors and their technologies differ; when industries are described as systems, the concept of sectoral systems of innovation is used to examine the development of individual sectors (Malerba, 2004), thereby capturing the particular techno-economic features and institutional structure, and concomitant changes. A specific research tradition within the sectoral approach centers on the concept of the industry life cycle (Abernathy and Utterback, 1978; Utterback, 1994). Accordingly, industries follow certain dynamics of innovation, firm entry into the market and growth, market structure, etc. In particular, the cycle is characterized by interdependent rates of product and process innovation over time which, in turn, are linked to significant transformations in the characteristics of products, processes, competition, and organization.

In addition to the more conventional analysis of evolutionary approaches, the focus is increasingly turning toward how technology and organizations within a sector develop simultaneously in a coevolutionary process with regard to their given environment (Van den Bergh and Stagl, 2003; Garnsey and McGlade, 2006; McKelvey, 1997). Coevolutionary processes involve knowledge, agents, demand, and institutions. Mutual learning between agents, positive network externalities, and an increasing carrying capacity may lead to an acceleration of diffusion of new technology in a self-propagating process (Nagamatsu et al., 2006). On the other hand, network externalities and increasing returns may also lead to path dependency and irreversible change that lock sectoral systems into inferior technologies (Arthur, 1994).

In a state-of-the art paper on evolutionary approaches, Malerba (2006) asserted that evolutionary studies cover a whole range of approaches, spanning from descriptive, historical analysis—including stylized facts—to longitudinal analysis identifying statistical regularities and statistical properties of change. In addition, models of industry dynamics with rational actors and models with bounded rational actors exist, alongside learning and experimentation, and trial and error. This book covers this range of variation characterizing evolutionary studies on innovation and industries. In terms of topic, this book deals with the diffusion of technology interaction within institutional systems and technology-based economic growth. It provides new insights from applied studies on information technology and biotechnology as sectors developing within an interactive framework involving different institutional systems and dynamics. State-of-the art results on learning processes in entrepreneurship are also presented, as well as innovation with a focus on the interface between science and business in different institutional systems.

Continuous innovation is essential for an economy, considering that an ongoing selection process takes place in each individual economic sector, as well as in the economy as a whole. In fact, an ongoing process of creative destruction is evident (in the Schumpeterian way of thinking) at the level of companies and, in the longer term, at the level of economic sectors. According to Rogers Everett (1962/1983), *diffusion* is the process by which an innovation is communicated through certain channels among the members of a social system. Studies on the diffusion of innovations range from microscopic views about the rationale behind the adoption of a new technology by innovators (such as leaders and followers) and their learning processes, to approaches that incorporate market variables, such as price and technology imports, to account for differences in diffusion. Diffusion on the aggregate level has been modeled using different growth functions, including the simple logistic growth function, the Gompertz curve, and the bi-logistic growth function (Meyer, 1994). In more recent models, innovations are perceived as interacting with and altering institutions while displaying systematic changes in their growth path. Such changes may derive from network externalities, as observed in the diffusion of information technology. This can lead to adoption rates based on a steadily increasing carrying capacity (Nagamatsu et al., 2006; Watanabe et al., 2003).

Diffusion is highly dependent on institutions. Institutions can be described as humanly devised constraints that shape interaction (North, 1990). Institutions reduce uncertainty by providing formal (e.g., rules and laws), as well as informal structures (such as conventions and codes of behavior). In economic relations, institutions reduce transaction costs, e.g., by circumventing contract and control costs. Typically, institutions change incrementally over time. Formal rules or regulations may, on the other hand, change overnight due to political or legal decisions; however, the associated informal institutions often remain unaffected for a long time, leading to periods of discontinuity in technological change. Institutions that influence innovation diffusion deal with intellectual property protection through patents and trademarks (i.e. formal aspects) and with learning practices (informal aspects), like conduct within learning networks through which diffusion takes place. Other crucial institutions deal with issues companies face in terms of technology investment, like behavior of venture capitalists (formal) and business angels (informal) (Edquist, 1997). Recent studies have dealt with the elasticity (or flexibility) of institutions. Due to information technology's close link with institutions, its distinctive functionality is influenced by a dynamic interaction with institutional systems. According to this coevolutionary view, institutional elasticity is a crucial factor in generating innovation's self-propagating behavior over time (Watanabe et al., 2004). This particular growth trajectory is explored in various chapters in Part I of the book.

From an evolutionary perspective, attention is increasingly turning to learning processes through which agents develop and adapt specific skills and capabilities, while operating in highly dynamic and uncertain market and technology environments. Learning firms feature different capabilities that enable them to build, integrate, and reconfigure internal and external competences, with networking being one of these capabilities (Cowan et al., 2004; Grant, 1996). In this context, the focus of analysis may shift to effective learning strategies and the interaction of firms displaying particular characteristics of the learning environment, including networking with public knowledge institutions like universities and institutions of higher education and research (D'Este and Patel, 2007). Universities, for their part, have become important players in efforts to commercialize new knowledge by contributing to incubators to foster spin-off companies (Shane, 2004). In fact, collaboration between universities, the business world, and governments has increased with the objective of reaping the fruits of knowledge commercialization. However, the formats may differ due to the different dynamics in national (and regional) institutions (Leydesdorff and Etzkowitz, 1998). These and similar themes are addressed in Part II of this book.

What the themes in the book share is their concentration on different institutional changes. The common framework of coevolutionary dynamics is elaborated in several chapters to elucidate differences over time and between countries in innovation and economic growth. The book builds on solid, applied research work, and reflects mainly on information technology and the electronics industry (for example, PC manufacturing), but also includes a chapter on biotechnology. Furthermore, the book has a wide geographic scope, covering countries such as the USA, China, Japan, India, and the Netherlands, and includes several cross-country comparisons as well (Table 1).

2 Content of the Book

The book is divided into two parts (Table 1). Part I (*Adoption of Information Technology in National Institutional Systems*) presents studies on historical development trajectories and/or country-specific institutional conditions of innovation diffusion. Special attention is paid to information technology as a separate sector (i.e. PC manufacturing) and as a supportive technology applied in other sectors (automotive industry). Part II (*Institutional Systems, Entrepreneurship, Knowledge Transfer and Learning*) is devoted to the development of institutions, policies, and practices of technology transfer and utilization by companies. Different stages in an evolutionary growth trajectory of technology incubators as a starting point for the establishment of new companies are analyzed. In addition, new forms of entrepreneurship and innovation are addressed from a perspective of learning and knowledge acquisition, as well as with reference to national policies in the context of institutional transformation.

Brief Summary of the Chapters

Part I Adoption of Information Technology (IT) in National Institutional Systems

In Chapter 2, *Lei, Kita, Watanabe* and *Tou* examine the development of China's PC manufacturing industry since the early 1980s. They focus on three different processes that have shaped this development, i.e., the effective utilization of spillover technology from imported PC components, accelerated growth of China's PC sector through market learning effects, and the creation of high functionality with the objective of catching up with international PC manufacturers. These three processes are evaluated and their influence on the development trajectory of China's PC industry determined. In addition, an international comparison of China with 15 developing PC-producing countries illustrates the shift to the phase of highly competitive production in accordance with global standards.

In Chapter 3, *Watanabe, Lei, Yamada* and *Kita* elaborate on the development trajectory of China's PC industry. In a cross-national comparative analysis, the rate of development of China's PC industry appears to be the highest in the world, and its institutional structure seems to be quite distinct in terms of national strategy/socio-economic system, entrepreneurial organization and culture, and historical perspectives. The authors highlight the influence of multiplier effects (phased development) in activating the PC sector's development in China, as well as the coevolution of activated PC development and institutional factors in accelerating innovation performance at the global level.

In Chapter 4, *Jauhari* explores the use of IT in India's automotive industry. India is emerging as a hub for the outsourcing of automotive components and vehicle

Table 1. Outline of the Book

Chapter and author	Theme	Research Approach and Methodology	Country
Chapter 1 (editors)	Overarching theories, short summary of contributions		
Part I	*Adoption of Information Technology in National Institutional Systems*		
Chapter 2 (Lei et al.)	Development trajectory of PC industry and decisive institutional factors	Empirical, explanatory analysis (hypothesis testing)	China, partially Japan, etc.
Chapter 3 (Watanabe et al.)	Coevolution of institutional structures, development of PC penetration and diffusion	Empirical, explanatory analysis (hypothesis testing)	China and partially other countries
Chapter 4 (Jauhari)	Development trajectory of vehicle manufacture and implementation of information technology	Empirical, descriptive analysis, including company case study	India
Chapter 5 (Griffy-Brown et al.)	Information technology and effectiveness of national institutional system shifts toward an information society	Empirical, descriptive analysis	USA, Japan, and China
Chapter 6 (Jauhari)	Factors determining export intensity of electronics firms	Empirical, explanatory analysis (hypothesis testing)	India
Part II	*Institutional Systems, Entrepreneurship, Knowledge Transfer and Learning*		
Chapter 7 (Van Geenhuizen & Soetanto)	Factors influencing growth of technology incubators, with an emphasis on evolutionary change	Empirical, explanatory analysis	Various countries across the world
Chapter 8 (Masurel & Van Werkhoven)	Challenges to the establishment of learning relations between SMEs and knowledge institutions	Empirical, descriptive analysis	The Netherlands
Chapter 9 (Van Geenhuizen)	Entrepreneurship changes in biotechnology and changing institutional transformations as a result of national policy	Empirical, descriptive analysis; policy analysis	The Netherlands and other European countries

manufacturing on account of its cost and quality advantages. Building on a value chain perspective, the study identifies the institutional context of India's passenger vehicle industry, followed by a case study of India's market leader in this sector, Maruti Udyog. The findings indicate that the use of IT is pervasive; for example, IT has made it possible to offer additional customer services, like vehicle financing, insurance, supply of accessories, etc. It has improved supply chain- and distribution management considerably, resulting in the supply of inventory in real-time. It is, however, only a matter of time until new organizational models are introduced, like the Modular Vehicle Platform, which will require new types of IT technology support.

In Chapter 5, *Griffy-Brown, Zhu* and *Jin* analyze institutional transformation in Japan, China, and the USA in a comparative way, with a focus on those changes that have facilitated the adoption of information technology (IT) by businesses to recast their relationships and operations. In a qualitative investigation, institutional changes in the three countries are explored and the constraints of such a comparison identified. In the USA, the system's rules are explicit and well-defined, and the institutional configuration is fluid, yet also constrained by these clearly outlined rules. IT corresponds well with such institutional characteristics, which, in turn, leads to flexibility. Japan's institutional system is very stable, but also less flexible and less transparent than the USA's. Overall, the utilization of IT in Japan seems one-sided, as it requires explicit institutional rules and open communication standards which are not well-developed. Consequently, the USA demonstrates widespread use of tools for its IT-driven economy, whereas Japan's use of IT in business processes is spreading to SMEs at a different pace. The case of China is difficult to compare with those of the other two countries, because institutional changes in China are not only distinct but occur on a different scale.

In Chapter 6, *Jauhari's* starting point is that exports can intensify developing economies' competitiveness. A review of the literature on factors that determine exports is complemented by the development of a model of export intensity, including firm size and various technology import indicators. By testing the model on electronic firms in India for the years 1990 to 2003, the results indicate that firm size, amount of foreign investment, and import of product components and capital equipment have a positive influence on export intensity. In an alternative model which merges the various technology import indicators into a single technology indicator, size and the technology indicator represent the determinants of the export intensity of India's electronics industry. This result calls for a reduction of tariff barriers for technology imports and the establishment of single window systems.

Part II Institutional Systems, Entrepreneurship, Knowledge Transfer and Learning

In Chapter 7, *Van Geenhuizen* and *Soetanto* examine the growth patterns of technology incubators as institutional constructs aimed at enhancing entrepreneurship, knowledge transfer, and learning. In the context of evolutionary change, three institutional characteristics are included in a growth model that explores type of stakeholder involvement (single versus multiple stakeholders), type of support (conventional versus value-added), and the incubator's overall growth strategy (focused versus broad and profit-oriented), in addition to various factors of the production system. Using a cross-national database, it appears that the combination of multiple stakeholder involvement and location in a non-metropolitan environment tends to foster a strong growth of incubators. The combination of conventional support and a focused strategy by the incubators also plays a role, but is linked to weak growth. Overall, the results demonstrate the positive influence of diversity and hybridization as outcomes of evolutionary processes.

Masurel and *Werkhoven* (Chapter 8) examine the role of public knowledge institutions with regard to learning by small- and medium-sized enterprises (SMEs) in the Zwolle region (Netherlands) within the context of a policy to speed up processes of innovation. The research elucidates that only a small minority of the SMEs surveyed had established partnerships with public knowledge institutions. The majority of enterprises that collaborate with knowledge institutes work together with universities of professional education. Firms that had no experience in collaborating with knowledge institutions frequently stated that they were unfamiliar with the services offered by such institutions and with the associated procedures. Firms that had worked with knowledge institutions, on the other hand, found the time-consuming nature of this type of interaction, as well as the unclear instructions by their partners to be problematic. The chapter suggests that learning processes between SMEs and public knowledge institutions may not be of significance for all SMEs, but are for those engaged in new technology development and in an early growth stage.

In Chapter 9, *Van Geenhuizen* elucidates the institutional constraints in biotechnology entrepreneurship in the Netherlands within an evolutionary context. A discussion of the national policy in the Netherlands is followed by an analysis of the biotechnology sector, particularly the importance of spatial clusters for innovation. This serves as input to a critical examination of the government's policy, together with actual changes taking place in the industry and within the relevant institutions. Two potential drawbacks of the government's policy are identified, i.e., the distribution of support measures among six different locations, thereby reducing the opportunity to create critical mass in one location, and the policy's emphasis on addressing a "fixed" number of start-ups, thus not taking full account of the strengths of such companies. Using a cross-national comparison, it becomes clear that academic entrepreneurship could significantly increase within a couple of years, if a comprehensive policy approach is implemented. However, the role of large domestic companies or equivalent structures in boosting the further growth of small start-ups cannot be enforced from above and takes time to develop. Also, institutional obstacles in financing start-ups' subsequent growth stages are difficult to remove. Nevertheless, the biotechnology sector in the Netherlands indicates positive signs of institutional transformation to overcome these barriers to growth.

References

Abernathy W J, Utterback J M (1978) Patterns of innovation in technology. Technology Review 80(7):1–47

Aghion P, Howitt P (1998) Endogenous growth theory. The MIT Press, Cambridge

Arthur W B (1994) Increasing returns and path dependency in the economy. The University of Michigan Press, Ann Arbor

Audretsch D B, Keilbach M C, Lehman E E (2006) Entrepreneurship and economic growth. Oxford University Press, Oxford

Cohen W M, Goto A, Nagata A, Nelson R R, Walsh P (2002) R&D spillovers, patents and the incentives to innovate in Japan and the United States. Research Policy 31(8–9):1349–1367

Cowan R, Jonard N, Ozman M (2004) Knowledge dynamics in a network industry. Technological Forecasting and Social Change 71(5):469–484

D'Este P, Patel P (2007) University–industry linkages in the UK: What are the factors underlying the variety of interactions with industry? Research Policy 36(9):1295–1313

Dosi G, Marsili O, Orsenigo L, Salvatore R (1995) Learning, market selection, and the evolution of industrial structures. Small Business Economics 7(6):411–436

Edquist C. (ed) (1997) Systems of innovation; technologies, institutions and organizations. Pinter, London

Garnsey E, McGlade J (eds) (2006) Complexity and co-evolution; continuity and change in socio-economic systems. Edward Elgar, Cheltenham

Grant R M (1996) Toward a knowledge-based theory of the firm. Strategic Management Journal 17:109–122

Grossman G M, Helpman E (1991) Innovation and growth in the global economy. The MIT Press, Cambridge

Hodgson G M (1999) Evolution and economics. On evolutionary economics and the evolution of economics. Edward Elgar, Cheltenham

Leydesdorff L, Etzkowitz H (1998) The Triple Helix as a model for innovation studies. Science and Public Policy 25:195–203

Malerba F (ed) (2004) Sectoral systems of innovation. Cambridge University Press, Cambridge

Malerba F (2006) Innovation and the evolution of industries. Journal of Evolutionary Economics 16(1/2):3–23

McKelvey M (1997) Coevolution in commercial genetic engineering. Industrial and Corporate Change 6(3):503–532

Meyer P S (1994) Bi-logistic growth. Technological Forecasting and Social Change 47(1) 89–102

Nagamatsu A, Watanabe C, Shum K L (2006) Diffusion trajectory of self-propagating innovations interacting with institutions—Incorporation of multi-factors learning function to model PV diffusion in Japan. Energy Policy 34:411–421

Nelson R R, Winter S G (1982) An evolutionary theory of economic change. Harvard University Press, Cambridge

North D C (1990) Institutions, institutional change and economic performance. Cambridge University Press, Cambridge

Rogers Everett M (1962, 1983) Diffusion of innovations. The Free Press, New York

Romer P (1990) Endogenous technological change. Journal of Political Economy 98(5)pt. 2:S74–S75

Shane S (2004) Academic entrepreneurship: University spinoffs and wealth creation. Edward Elgar, Cheltenham

Utterback J M (1994) Mastering the dynamics of innovation. Harvard Business School Press, Boston

Van den Bergh J, Stagl S (2003) Coevolution of economic behaviour and institutions: Towards a theory of institutional change. Journal of Evolutionary Economics 13(3):289–317

Watanabe C, Kondo R, Ouchi N, Wei H (2003) Formation of IT features through interaction with institutional systems—Empirical evidence of unique epidemic behaviour. Technovation 23(3):205–219

Watanabe C, Kondo R, Ouchi N, Wei H, Griffy-Brown C (2004) Institutional elasticity as a significant driver of IT functionality development. Technological Forecasting and Social Change 71:723–750

Part I
Applied Studies: Adoption of Information Technology in National Institutional Systems

An Empirical Analysis of the Institutional System's Effects on the Development of China's Personal Computer Industry—From Inertia to Innovation

Shanyu Lei, Masanobu Kita, Chihiro Watanabe, and Yuji Tou

Abstract In line with the global advancement of information technology (IT), China's personal computer (PC) industry has undergone a dramatic development in recent years. Worldwide, China ranked fourth in 2002 in total number of PCs installed, following the USA, Japan, and Germany. This marks significant progress, considering that China only ranked 16th in 1993. This leap can be attributed to China's annual growth rate; at 42.9%, China is the frontrunner in the world. The quality of China's PCs has also improved considerably. As a result, total sales by indigenous PC manufacturers caught up with those by international PC makers in 1997. Furthermore, Lenovo, China's largest PC manufacturer, purchased IBM's PC division in 2004, increasing the company's world market share from only 2% (rank 9 in the world) to 7.6% (rank 3). The surge in China's PC production suggests the following hypothetical reasons for the PC industry's evolution from initial inertia along a continuously rising trajectory to its current state of success: (i) effective exploitation of technology spillover from imported PC components; (ii) the acceleration of market learning effects; and (iii) a shift toward high functionality development, enabling Chinese PC makers to catch up with international counterparts in the highly competitive market environment of the late 1990s. This chapter attempts to verify the aforementioned hypotheses through an empirical analysis of the factors that have shaped the development of China's PC industry.

Key words: Institutional Systems, Technology Spillover, Market Learning, Initial Inertia

---

Corresponding Author: Chihiro Watanabe
Department of Industrial Engineering & Management, Tokyo Institute of Technology, 2-12-1, W9-49, Ookayama, Meguro-ku, Tokyo 152-8552 Japan, e-mail: chihiro@me.titech.ac.jp

M. V. Van Geenhuizen et al., (eds). *Technological Innovation Across Nations*,
DOI: 10.1007/978-3-642-00158-1_1,

1 Introduction

1.1 The Effect of Initial Inertia on the Resulting Trajectory of China's PC Industry

With the rapid development of the world economy, China's consumer electronics and durable consumer goods industries have undergone a dramatic development. The world has focused in particular on the remarkable transformation of China's PC industry, which grew seven times faster than the color TV and automobile sectors in the period 1995–2002. China's global rank in terms of total PCs installed improved considerably from 16th in 1993 to fourth in 2002, signifying a substantial increase in PC demand. In response to this dramatic rise in demand, China's indigenous PC manufacturers undertook numerous efforts to improve the quality of their PCs. Looking back at the development trajectory of China's PC industry and the major milestones of the last two decades, two critical turning points emerge. The first one came in 1992, when foreign capital investment was deregulated. The second turning point was in 1997, when the sales of indigenous PC manufacturers exceeded those of international counterparts. The initial inertia characterizing China's PC sector can thus be classified into two periods: *an imported parts-dependent period* (1981–1992) and *a market expansion period* (1993–1997). The industry's initial inertia evolved into a steadily rising trajectory, resulting in noteworthy achievements during *the competitive market period* (after 1998).

China's distinct institutional system is directly linked to the PC industry's development; institutional systems are defined as three-dimensional systems, consisting of (i) national strategy and the socio-economic system, (ii) entrepreneurial organization and culture, and (iii) historical perspectives (Watanabe, 2005). China's national strategy and socio-economic system serve as the focal point in our analysis of the PC sector's development trajectory.

1.2 Sources of High Performance in China during the PC Industry's Initial Inertia

1.2.1 Imported Parts-Dependent Period (1981–1992)

Trends in imported PC components and number of PCs produced in China from 1981 to 1997 are illustrated in Fig. 1. The parallel path between the two during the imported parts-dependent period shows that China's PC manufacturers were completely dependent on imported parts.[1] This suggests that the development of China's

[1] During this period, China's economic system evolved from a planned to a market-oriented economy, also referred to as a 'social market economy with Chinese character.' The Chinese government aimed to achieve economic growth by increasing imports from abroad, through which knowledge spillovers could be absorbed.

PC technology relied fully on technology spillover from exporting countries during this period.

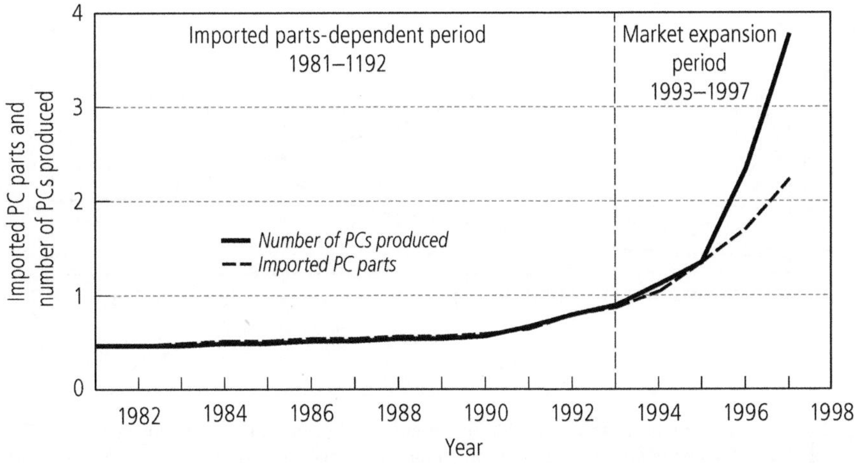

Fig. 1. Trends in Imported PC Parts and Number of PCs Produced in China (1981–1997); Index: 1995=1
Source: China Industrial Economy Statistical Yearbook, and China Customs Statistical Yearbook

1.2.2 Market Expansion Period (1993–1997)

From the mid-1990s, the number of PCs produced in China exceeded that of imported PC parts, indicating that the dependency on imported parts decreased. Fig. 2 compares trends in the number of PCs per 1000 inhabitants and prices of PCs in China over the period 1988–1997. This figure clearly illustrates that as the diffusion of PCs sharply increased, their price dropped, even in the period following the decline of PC manufacturers' dependence on imported components. This implies that China fully benefited from market learning during the course of the PC sector's development and subsequent deployment process, particularly in the period of market expansion, when dependency on imported PC components began to decrease.

1.2.3 Competitive Market Period (After 1998)

Fig. 3 compares PC sales and prices of indigenous and international manufacturers in China from 1993 to 2001.

This figure reveals that the sales of indigenous PCs soared in the competitive market period (after 1998), after surpassing those of foreign producers in 1997. Corresponding to this increase, the price gap between indigenous and international

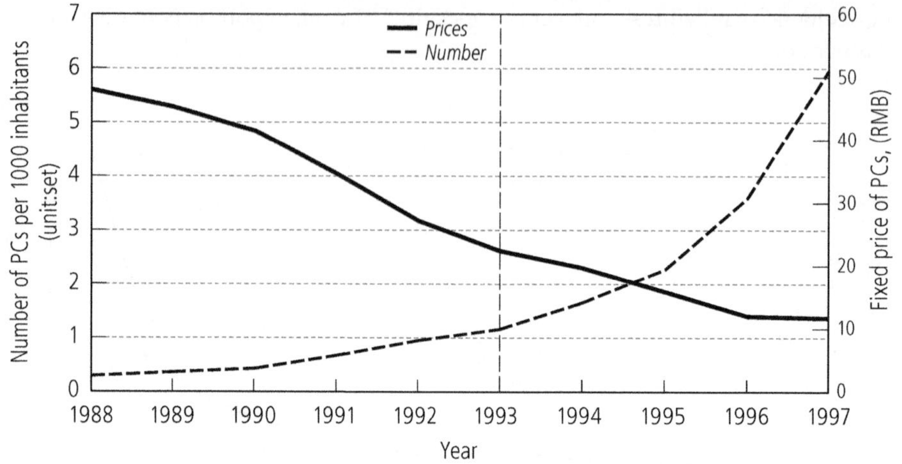

Fig. 2. Trends in Number of PCs per 1000 Inhabitants and Price of PCs in China (1988–1997)
Data source: Chinese Ministry of Electronic Industry Materials and World Telecommunication Indicators Database 2004 (International Telecommunication Union (ITU))

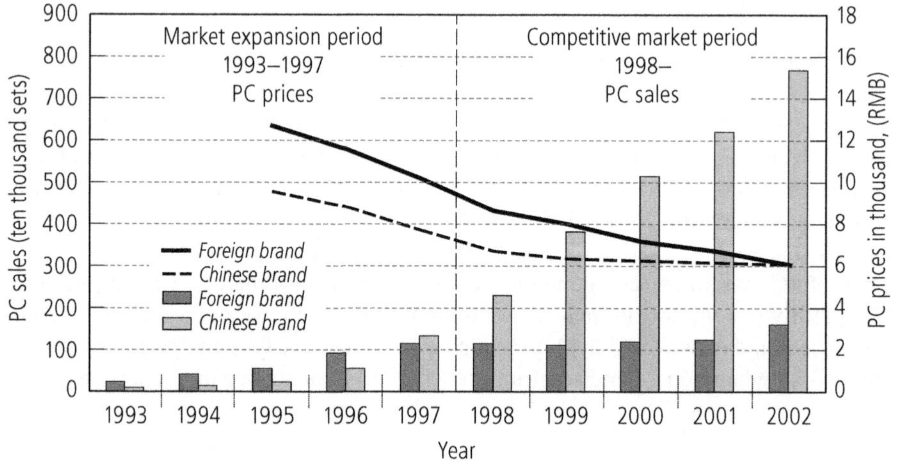

Fig. 3. Comparison of PC Sales and Prices of National and International Manufacturers in China (1993–2001)
PC prices of Chinese brands are based on a three-year average.
Data source: Chinese Ministry of Electronic Industry Materials and China Computer Newspaper

brands diminished until the price of Chinese brands exceeded that of foreign ones in 2002. This can be attributed to efforts by indigenous PC makers to create high functionality, with the aim of catching up with foreign products within the competitive market.

1.2.4 Hypotheses

The previous observations have prompted the hypothetical assumption that the initial inertia of China's PC industry shaped the resulting upward trajectory, thus generating its notable success. This can be attributed to the following institutional and technological advancements:

1. Effective utilization of technology spillover from imported PC parts during *the imported parts-dependent period* (1981–1992);
2. Acceleration via market learning effects during *the market expansion period* (1993–1997); and
3. Intensive efforts to create high functionality corresponding to the shift toward a competitive environment in *the competitive market period* (after 1998).

1.3 Previous Research

1.3.1 Technology Spillover

As Griliches (1979) contends, firms' level of technology knowledge depends not only on internal R&D investment, but also on the knowledge stock developed by other firms or industries. These external effects are referred to as technology spillover. The increasing significance of technology spillover for firms striving to remain competitive has been corroborated by several studies (OECD, 1997, 1998; Watanabe et al., 2003, 2004). International technology spillover occurs through the import or absorption of foreign direct investment (Coe and Helpman, 1995; Keller, 1996). Coe and Helpman (1995) also contend that the direct benefits of international R&D include learning about new technologies, production processes, and organizational methods. Lumenga-Neso et al. (2005) provided empirical results indicating that indirect trade-related R&D spillover is 14 times greater than direct spillover with regard to its contribution to total factor productivity. This implies that indirect trade-related spillover is as important as direct trade. Competitive pressure from international firms induces national ones to increase their R&D expenditures and thus adapt to the intense competition characterizing the international market. This competition, in turn, accelerates technology spillover (Holmes and Schmitz, 2001; Lai et al., 2006).

Building on such preceding research, this study analyzes the significance of technology spillover in the accumulation of skills necessary to install imported components into a product, and to effectively utilize imported products.

1.3.2 Learning Effects

Previous research has indicated that experience, knowledge, or insight gained from the use of imported products has played a significant role in improving innovation diffusion (Venkatraman, 1988; Gatignon and Robertson, 1991). The rate of learning through complex products or special usage methods is quite remarkable. However, Jovanovic and Lach (1989) point out that the process of diffusion slows down as learning effects rise.

Learning by doing is a phenomenon that begins the moment new knowledge is initially implemented, and lasts throughout the maturation stage, providing opportunities for cost reduction and quality improvements (Gregory, 2006). Learning also plays an important role in generating new technological knowledge and the introduction of new technologies (Antonelli, 2006).

According to Abernathy (1978), the initial innovation level of PCs built on former technology—the dynamics underlying its origin and realization depended not only on the manufacturer, but on the economic and technological environment, as well as on market structure. That is, market learning can be defined as a series of learning processes through production, distribution, and utilization. Watanabe and Asgari (2004) developed a mathematical model to measure dynamic learning coefficients representing market learning effects, and successfully tested the model's usefulness by determining the diffusion trajectory of Japan's photovoltaic cells industry.

Checkland (1988) asserts that the conduct of business and certain decisions are subconsciously influenced by historical experience, i.e., by cumulative learning, and emphasized the significance of the mind's 'historical stock-taking'. Checkland (1985) also stressed the influence of consciously progressing "from optimizing to learning", which suggests that market learning is far more effective than purely theoretical solutions. Building on Checkland's findings, we can conclude that market learning accelerates the optimal shift in business resources, which, in turn, further enhances market learning. That is, it is possible to construct a mutually beneficial cycle.

Our study rests on these groundbreaking ideas to underline the key role spillover technology played during the imported parts-dependent period, its function as a catalyst to advance market learning throughout the market expansion period, as well as generating high functionality development during the competitive market period. Consequently, individual technology policies are required for the different stages of innovation (Cantner and Pyka, 2001).

1.3.3 Technology Diffusion

Earlier studies generally describe the diffusion pattern as following a sigmoid (S-shaped) curve (Verhulst, 1845; Griliches, 1957). Gomulka (1971) was the first to point out that price competition could be represented by a sigmoid curve corresponding to economic development over time. In other words, the rate of competition increases from a low to a high level, decreases after reaching the inflection point to finally converge within a stable environment, once perfect competition has been achieved. Based on Gomulka's work, we computed technological competition by using a logistic function to analyze the diffusion of technology in China's PC sector, taking competition from international PC manufacturers into account.

Furthermore, some noteworthy research linked to technology diffusion, namely network externality, inspired our work. Katz and Shapiro (1985) claim that hardware sales in the PC industry depended on the expansion of complementary products, i.e., software and its capabilities. The network externalities of hardware and software indirectly influence the sale of PCs, since network externality implies that the benefits a given product provides increase in correspondence with the scale of the given network (Oren and Smith, 1981). Thus, it is essential to exceed critical mass when network externality emerges, as the number of customers increases considerably through self-propagation. Compared with other goods in general, the diffusion curve of PCs increases gradually before reaching the critical mass point, and sharply rises and diffuses thereafter, at accelerated speed (Oren and Smith, 1981; Rogers, 1983).

Prompted by previous research and the premise that the upward trajectory and current success of China's PC industry can be attributed to its initial inertia, our study attempts to empirically analyze our hypothetical assumptions.

Section 2 presents the analytical framework. A comparative empirical analysis of the PC industry's evolution in China and Japan[2] is carried out in Section 3, and the findings, along with policy implications and recommendations for future study, are provided in Section 4.

2 Analytical Framework: Trajectory of PC Development and Diffusion

2.1 Dynamic Learning Coefficients and Diffusion of PCs

The function describing market learning in the course of PC diffusion can be depicted as follows:

$$P = A \cdot PC_T^{-\lambda} \tag{1}$$

where P is fixed price of PCs; A is scale factor; PC_T is cumulative number of PCs; and λ is market learning coefficient.

Since the market learning coefficient λ is a function describing successive coefficients during the diffusion process' production, distribution, and utilization phase, these coefficients can be depicted as a function of time trend t:

[2] Japan is included in the analysis, since China predominantly imports PC parts from Japan, and began absorbing more investment from Japan following the opening of its economy. Furthermore, the authors, who were involved in the research study, are Japanese and Chinese.

$$\lambda(t) = \lambda(\lambda_1(t), \lambda_2(t), \ldots, \lambda_n(t)) \approx \sum_{i=0}^{n} a_i n^i \tag{2}$$

where a_i is the ith coefficient of time trend t.

The dynamic learning coefficients $(\lambda_1(t), \lambda_2(t) \ldots \lambda_n$, represent market learning effects in a successive production, distribution, and utilization process. If $\lambda_{(t)}$—depicted in equation (2)—is substituted for λ—depicted in equation (1)—in a logarithm application, the following equation emerges:

$$\ln P = \ln A - \sum_{i=0}^{n} a_i t^i \ln PC_T + \varepsilon \tag{3}$$

where ε is a disturbance term which is independent from $\lambda(t)$.

Taking the differentiation of equation (3) $\ln PC_T, \lambda(t)$ can be determined.[3]

2.2 Price Decline Resulting from the Import of PC Parts and International Competition

The disturbance term in equation (3) ε includes (i) spillover effects from imported PC components, and (ii) the influence of decreasing prices resulting from competition with international manufacturers. Defining elasticity of imported PCs to price as $\delta(t)$, and the influence of competition with international manufacturers on price as $D(t)$, the disturbance term can be depicted as follows:

$$\varepsilon = \delta(t) \ln IM + cD(t) \tag{4}$$

where IM is imported PCs; and c is coefficient.

Similar to the learning coefficient, the elasticity of imported PCs to price $\delta(t)$ can be depicted as a function of time trend t:

[3] The relationship between fixed prices of PCs P and cumulative number of PCs PC_T can be depicted as equation (3):

$$\ln P = \ln A - \sum_{i=0}^{n} a_i t^i \ln PC_T + \varepsilon$$

where ε is the disturbance term which is independent from $\lambda(t)$. Taking a differentiated equation (3) by $\ln PC_T$, learning coefficient $\lambda(t)$ can be computed as follows:

$$\frac{-\partial \ln P}{\partial \ln PC_T} \approx \sum_{i=0}^{n} a_i t^i + \ln PC_T \frac{\partial \sum_{i=0}^{n} a_i t^i}{\partial \ln PC_T},$$

and

$$a_i t^i = a_0 + \varepsilon(\varepsilon \ll a_0), \text{thus}, \frac{-\partial \ln P}{\partial \ln PC_T} \approx \sum_{i=0}^{n} a_i t^i = \lambda(t).$$

Therefore, the elasticity of imported parts as compared to price and elasticity of PCs to GDP $\eta(t)$ can be calculated using the same equation.

$$\delta(t) = \sum_{i=0}^{n} b_i t^i \tag{5}$$

where b_1 is the *ith* coefficient of time trend t.

In keeping with Gomulka (1971), since price competition increases in a sigmoid manner in correspondence with economic development, which also increases over time, the function of price competition $D(t)$ can be depicted by the logistic function:

$$D(t) = \frac{1}{1+e^{-at-b}} \tag{6}$$

where a and b are coefficients.

$D(t) = 0$ when there is no price competition, and $D(t) = 1$ when price competition is highest. Coefficients a and b in equation (6) can be identified by means of the following empirical evidence:

1. The inflection point $(-b/a)$ is reflected in the structural change of 1991, leading to the deregulation of foreign capital investment in 1992.
2. As depicted in Fig. 1, the dependency of China's PC industry on imported parts decreased rapidly after 1997. According to existing research, imports from abroad are predominantly responsible for technology spillover (Coe and Helpman, 1995; Keller, 1996). The spillover effect from imported PC parts decreased considerably after 1997, owing to the sharp decline in PC component imports when the number of PCs manufactured in China exceeded that of international PC makers, as illustrated in Fig. 3.

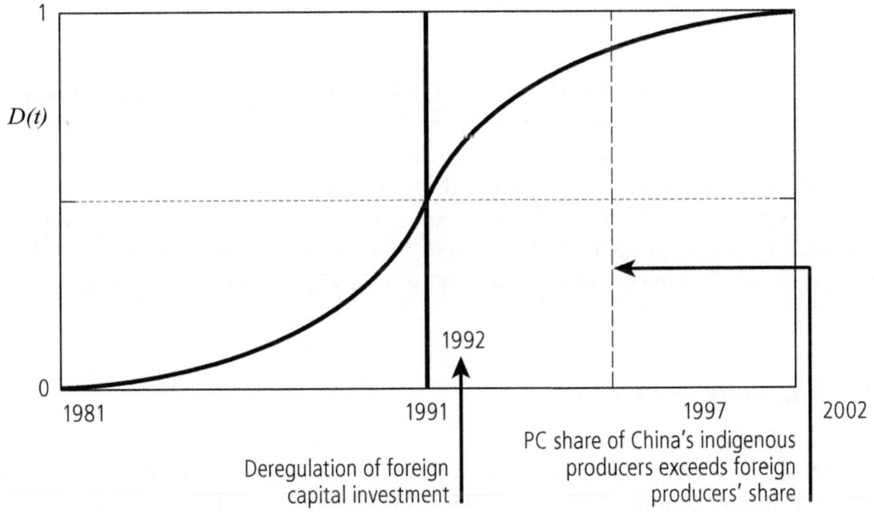

Fig. 4. Trajectory of PC Price Competition in China

Taking these factors into account, the function of PC price can be described by the following equation:

$$\begin{aligned}\ln P &= \ln A - \sum_{i=0}^{n} a_i t^i \ln PC_T + \varepsilon \\ &= \ln A - \sum_{i=0}^{n} a_i t^i \ln PC_T + \sum_{i=0}^{n} b_i t^i \ln IM + c\frac{1}{1+e^{-at-b}}\end{aligned} \tag{7}$$

2.3 Spillover Effect from Imported PC Parts

As suggested by growth accounting theory, a decrease in price signifies technological progress (Denison, 1962; Jorgenson and Griliches, 1967). Thus, the elasticity of imported PCs to price indicates the effect of technology spillover and is represented by the following equation:

$$\delta(t) = \frac{\partial \ln P}{\partial \ln IM} \approx \sum_{i=0}^{n} b_i t^i \tag{8}$$

where b_i is the ith coefficient of time t.

When $\delta(t) < 0$, the effect of technology spillover is evident, while it is not when $\delta(t) > 0$.

2.4 Influence of Price Competition

China's PC industry was established in 1981 and faced no price competition, i.e., $D(t_{1981}) = 0$ in equation (6). Given that $t = 0$ in 1981, the inflection point $(-b/a)$ in equation (6) was reached at time $t = 10$, namely in 1991, shortly before foreign capital investment was deregulated. In addition, since the effect of technology spillover came to an end in 1997 when the number of PCs sold by China's indigenous PC manufacturers caught up with the number of PCs being sold by international PC makers, $\delta(16) \geq 0$ in equation (8), as $t = 16$ in 1997, corresponding to $t_{1981} = 0$.

2.5 Elasticity of PCs to GDP

Similar to the increase in income, the diffusion of PCs also increases over time in "PC-developing countries" (Janes, 1995). Figs. 5 and 6 indicate trends in GDP per capita and number of PCs per 1000 inhabitants in PC-developing countries[4] from 1981 to 2002. In these two figures it becomes apparent that both GDP per capita and the number of PCs per 1000 inhabitants increase in accordance with time t. Figs. 5 and 6 also reveal the correlation analysis in the case of China, demonstrating the particularly strong statistical significance both in GDP per capita and number of PCs per 1000 inhabitants.

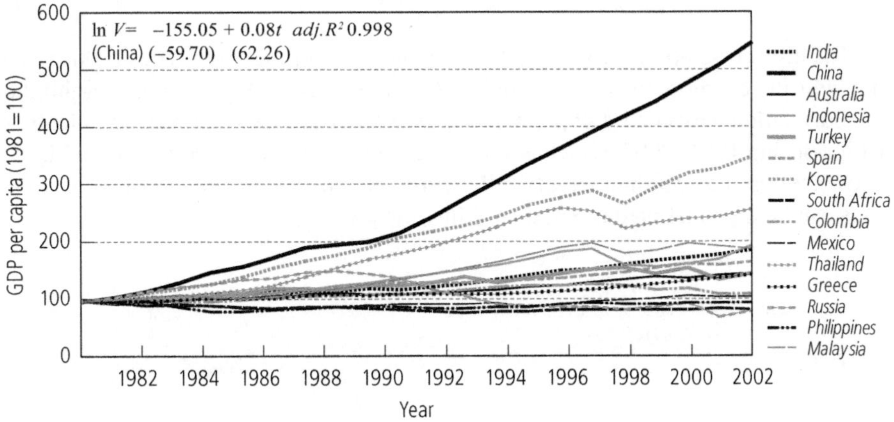

Fig. 5. Trends in GDP per Capita in PC-Developing Countries (1981–2002)

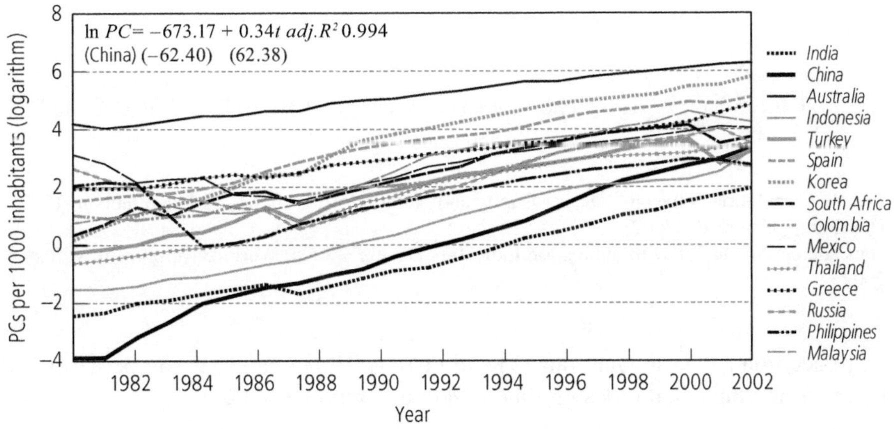

Fig. 6. Trends in Number of PCs per 1000 Inhabitants in PC-Developing Countries (1981–2002)
Source: World Telecommunication Indicators Database 2004; World Development Indicators 2002

[4] PC-developing countries are defined as countries whose PC industry is in an early stage of development and has not yet reached the inflection point in the PC development trajectory.

Based on these observations, the relationship between GDP per capita, as well as the diffusion of PCs and time t can be described as:

$$V = F_1(t), PC = F_2(t) \tag{9}$$

By synchronizing two equations, the diffusion of PCs can be depicted as a function of GDP per capita V:

$$PC = F_3(V) \tag{10}$$

The high correlation between GDP per capita and number of PCs per 1000 inhabitants in PC-developing countries in 2002 is illustrated in Fig. 7. Fig. 8 highlights the correlation between GDP per capita and number of PCs per 1000 inhabitants in China during 1981–2002, which is also extremely strong. The results of both Figs. 7 and 8 support the postulation that the PC diffusion trajectory in PC-developing countries can be demonstrated in relation to income level.

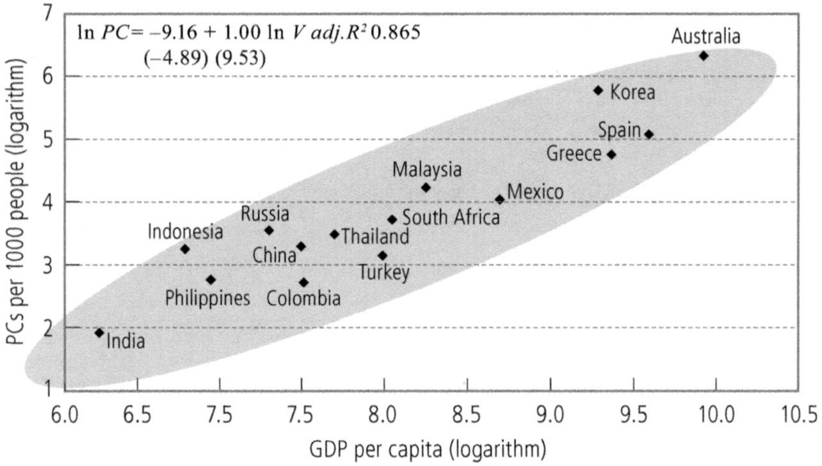

Fig. 7. Correlation between GDP per Capita and Number of PCs per 1000 Inhabitants in 15 PC-Developing Countries (2002)
Data source: World Telecommunication Indicators Database 2004; World Development Indicators 2002

In accordance with equations (9) and (10), *GDP* per capita V can be described by the following equation using time trend t and number of PCs:

$$V = F(t, PC) = F(PC(t)) \tag{11}$$

By introducing elasticity of PCs to $V(\eta)$, V can be depicted by the Cobb-Douglas type function:

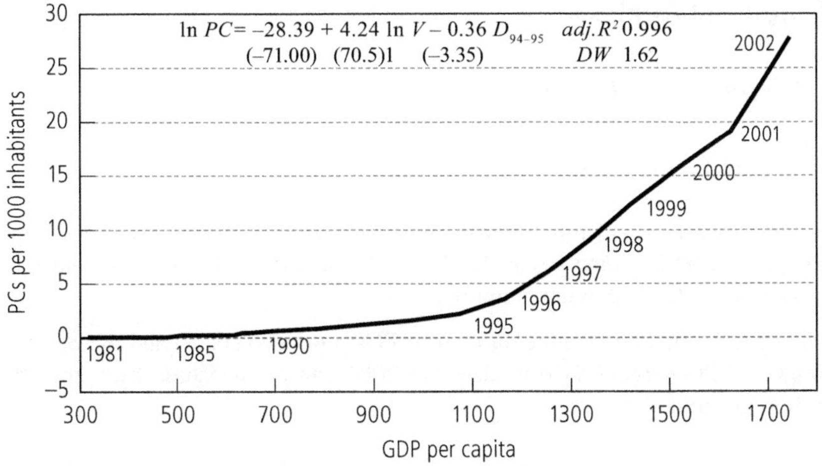

Fig. 8. Correlation between GDP per Capita and Number of PCs per 1000 Inhabitants in China (1981–2002)
Data source: World Telecommunication Indicators Database 2004; World Development Indicators 2002

$$V = B \cdot PC^{\eta} \tag{12}$$

where B is scale factor.

Similar to the dynamic learning coefficient, η, each phrase can be expressed as a function of t by:

$$\eta(t) = \sum_{i=0}^{n} c_i t^i \tag{13}$$

where c_i is the *ith* coefficient of column t.

Using the logarithm of equation (12) and substituting equation (13) for η, GDP per capita V can be depicted as:

$$\ln V = \ln B + \sum_{i=0}^{n} c_i t^i \ln PC \tag{14}$$

Taking partial differentiation of equation (14) by *lnPC*, the elasticity of PCs to GDP per capita can be described by:

$$\frac{\partial \ln V}{\partial \ln PC} \approx \sum_{i=0}^{n} c_i t^i = \eta(t) \tag{15}$$

3 Empirical Analysis

3.1 Trends in PC Prices

3.1.1 Trends in Fixed PC Prices in China

Since no consistent and reliable statistical data on PC prices (*P*) and on imports of PC parts (*IM*) in China are available for the period 1981–1990, they have been estimated using the following method[5]:

1. The correlation between the number of PCs produced (*PCP*) and the number of imported PC parts (*IM*) from 1991 to 2001 can be measured by applying the following equation:

$$\ln IM = -1.07 + 0.80 \ln PCP \quad \text{adj.}\ R^2 0.965 \quad DW \quad 1.04$$
$$(-4.09)(15.82) \tag{16}$$

2. Based on this result, the number of imported PC parts (*IM*) in the period 1981–1990 can be estimated by using the available data on the number of PCs produced (*PCP*) within that timeframe.
3. The correlation between number of PCs produced (*PCP*), of imported PC parts (*IM*), and price of PCs (*P*) from 1991 to 2001 can be measured by applying the following equation:

$$\ln P = 11.23 - 0.14 \ln PCP - 0.36 \ln IM \quad \text{adj.}\ R^2 0.965 \quad DW \quad 2.68$$
$$(60.86)(-1.33) \quad (-2.45) \tag{17}$$

4. The price of PCs in the period 1981–1990 can be estimated by using the data on *IM* and *PCP*.

Based on our measurements, the trend of fixed PC prices in China from 1981 to 2001 is illustrated in Fig. 9, indicating that the price of PCs has been decreasing consistently.

3.1.2 Trends in Fixed PC Prices in Japan

Japan's PC prices are measured using the same approach as applied to the case of China, namely by the following equation:

---

[5] Estimation results of equations (16) and (17) with a high value of adjusted R square indicates a high statistical significance. PC price and number of imported PC parts in the period 1981–1990 are computed based on the results with the least error. In addition, due to China's lack of reliable statistical data sources, the authors confirmed the estimated price by carrying out interviews in China, and found the estimated price to be reliable.

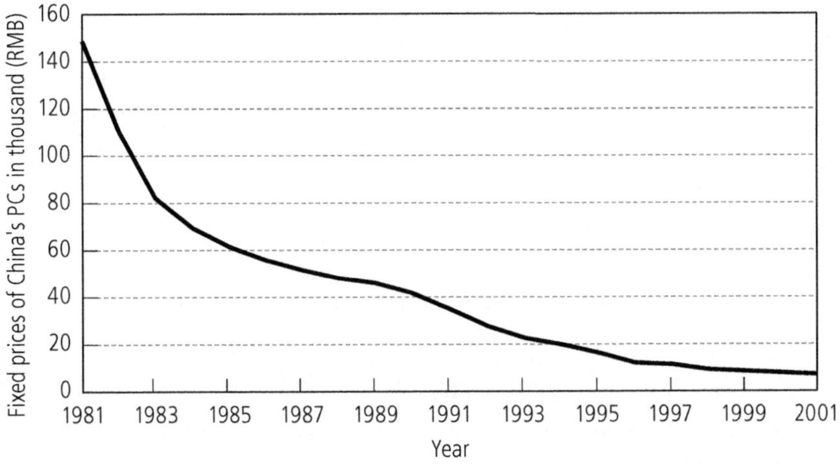

Fig. 9. Trend in Fixed PC Prices in China (1981–2001)
Data source: China Computer Newspaper and China Computer Education Newspaper

$$P = A \cdot PC_T^{-\lambda(t)} \tag{18}$$

where PC_T is the total installment of PCs in Japan (ten thousand sets); $\lambda(t)$ is the dynamic learning coefficient; P refers to fixed price of Japan's PCs (index: price in 2000); and A is scale factor.

Elasticity of PC production to number of sales by the eight leading PC producers[6] in Japan can be regarded as a proxy variable for the fixed PC price in Japan, and the sales of PCs S can be described by a function, which includes the number of PCs PCP and technology stock T[7]:

$$S = S(PCP, T) \tag{19}$$

Taylor expansion of equation (19) to the secondary term with regard to $\ln T$ and $\ln PCP$:

$$\ln S = d_0 + d_1 \ln T + d_2 \ln PCP + 1/2[\ln T \cdot d_3' \ln PCP + \ln PCP d_3'' \ln T]$$

[6] PCs manufactured by Matsushita, NEC, Hitachi, Toshiba, Fujitsu, Mitsubishi, Sony, and Sharp accounted for 73% of Japan's total PC production in 2002.

[7] Here, technology stock implies technological knowledge stock generated by R&D investment and—in line with the preceding approach (Griliches, 1979; Watanabe, 1999)—it can be measured with the equation:

$$T_t = R_{t-m} + (1-\rho)T_{t-1}$$

where T_t is technology stock at time t; R_t is R&D investment at time t; m the time elapsed between R&D and commercialization; and ρ is rate of obsolescence of technology.

$$= d_0 + d_1 \ln T + d_2 \ln PCP + d_3 \ln T \ln PCP \tag{20}$$

where S is total sales of Japan's eight major PC manufacturers; T is total technology stock of those eight PC makers; PCP is total number of PCs produced in Japan (unit: ten thousand sets); $d_i (i = 0\ 3)$ refers to coefficients, and $d_3 = 1/2(d_3' + d_3'')$.

The elasticity of PC production to sales can be depicted as:

$$\frac{\partial \ln S}{\partial \ln PCP} = \frac{\partial S}{\partial PCP} \cdot \frac{PCP}{S} = d_2 + d_3 \cdot \ln T \tag{21}$$

Consequently, marginal productivity of PCs can be expressed as:

$$\frac{\partial S}{\partial PCP} = \frac{S}{PCP}(d_2 + d_3 \cdot \ln T) \tag{22}$$

The results of the regression analysis of equation (20) from 1982 to 2002 can be enumerated as follows:

$$\begin{aligned} \ln S &= 0.66 + 0.94 \ln T + 0.54 \ln PCP - 0.05 \ln T \cdot \ln PCP \\ &\quad -0.14D \quad \text{adj.}\, R^2 \quad 0.995 \quad DW 1.88 \\ &\quad (0.79)(10.11) \quad (3.13) \quad (-2.90) \quad (-6.04) \end{aligned} \tag{23}$$

Provided that Japan's PC prices are determined by competition, their fixed price ought to equal their marginal productivity. By applying the coefficients identified in equation (23), the trend in Japan's PC prices can be determined and is illustrated in Fig. 10. The figure shows that while the price for PCs in Japan also consistently decreased, it began to stagnate in the latter half of the 1990s, implying that the development of Japan's PC industry slowed down in contrast to China's, because PC diffusion in Japan had reached the inflection point. In China, on the other hand, demand for PCs increased rapidly and PC diffusion is currently in a dynamic phase, approaching the inflection point.

3.2 Dynamic Learning Coefficient in China's PC Technology

By implementing the estimated price of PCs, the dynamic learning coefficient $\lambda(t)$ in the development of China's PC technology from 1981 to 2002 can be measured using the following equation:

$$\ln P = \ln A - \lambda(t) \ln PC_T + dD_{81} = \ln A - \sum_{i=0}^{n} a_i t^i \ln PC_T + dD_{81} \tag{24}$$

where D_{81} is a dummy variable (1981=1, other years=0).

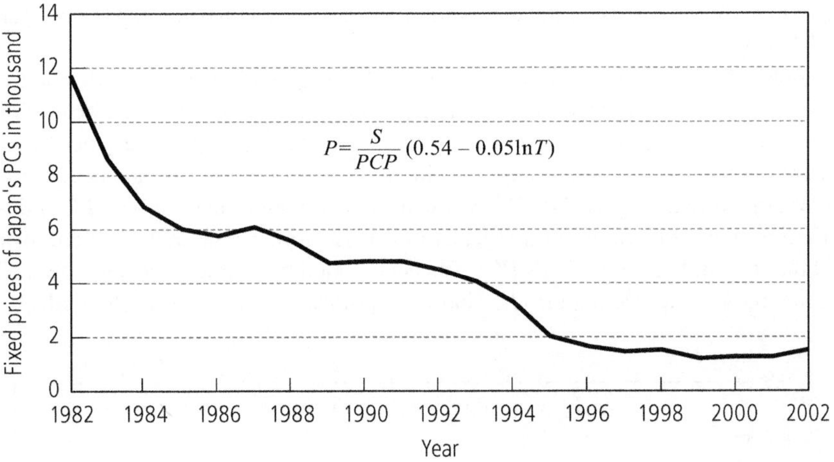

Fig. 10. Trend in Fixed PC Prices in Japan (1982–2002); Index: 1983=10

Table 1 summarizes the analytical results of the dynamic learning coefficient in the period 1981–2002. By comparing AIC (Akaike Information Criteria),[8] n=4 is identified as statistically most significant.

Table 1. Comparison of Dynamic Learning Coefficients of China's PC Technology (1981–2002)

	$\ln A$	a_0	a_1	a_2	a_3	a_4	a_5	d	adj. R^2	DW	AIC
$n=3$	11.79	0.37	–0.029	$3.2*10^{-3}$	$-8.7*10^{-5}$			0.30	0.996	1.38	– 51.79
	(3.37)	(3.63)	(–1.70)	(2.79)	(3.35)			(3.37)			
$n=4$	12.01	0.84 7	–0.17	0.020	$-9.3*10^{-4}$	$1.5*10^{-5}$		0.27	0.997	2.03	–62.76
	(122.99)	(5.35)	(–3.98)	(4.07)	(–3.82)	(3.47)		(3.93)			
$n=5$	11.89	–0.50	0.033	$-3.3*10^{-3}$	$-9.5*10^{-4}$	$5.7*10^{-5}$	$1.0*10^{-6}$	0.26	0.997	1.29	–62.65
	(85.61)	(1.48)	(0.26)	(0.15)	(–0.56)	(0.88)	(1.12)	(3.85)			

Data source: World Telecommunication Indicators Database 2004 (ITU); World Development Indicators 2002; China Computer Newspaper; China Computer Education Newspaper

The trend in the dynamic learning coefficient of China's PC technology from 1982 to 2002 is illustrated in Fig. 11. The figure reveals that the coefficient's value decreased in the 1980s, but increased again at the end of the 1980s and throughout

[8] The AIC is an operational method of 'trading off' the complexity of an estimated model against how well the model fits the data, and can generally be calculated using the following equation:

$$AIC = n \ln (RSS/n) + 2K$$

where n is the number of observations; RSS is residual sums of square; and K is the number of parameters in the model.

The preferred model is the one with the lowest AIC value.

the 1990s. The increase rate was strongest during the market expansion period (1993–1997).

Looking back at the possible effect of technology spillover on the development of China's PC industry as initially anticipated (Fig. 1), we see that the effect of technology spillover decreased from the 1990s onward, and that, simultaneously, the dynamic learning coefficient began to increase. This correlation has been verified by Watanabe and Asgari (2004).[9] We can thus conclude that China's PC industry shifted to market learning with regard to PC diffusion. These analyses demonstrate that the 'technology of China's PCs was accelerated by market learning effects during the market expansion period.' That is, hypothesis 2 can be substantiated.

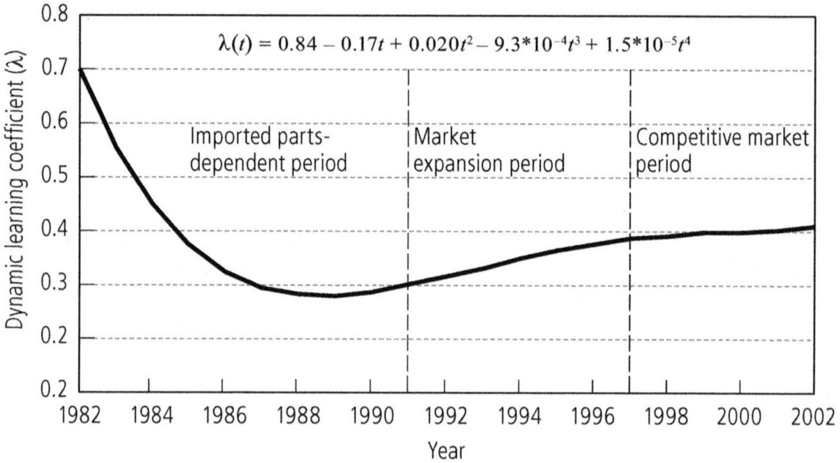

Fig. 11. Trend in Dynamic Learning Coefficient in China's PC Technology (1982–2002)

Similar to the measurement of the dynamic learning coefficients in China's PC technology, Japan's dynamic learning coefficients are determined in equation (24), and the analytical results for the period 1982–2002 summarized in Table 2.

$$\ln P = \ln A - \lambda(t) \ln PC_T + dD_{81} = \ln A - \sum_{i=0}^{n} a_i t^i \ln PC_T + dD_{82} \qquad (24')$$

By comparing AIC, $n = 3$ is identified as statistically most significant with regard to the learning coefficient in the development of Japan's PC technology. Fig. 12 illustrates this trend.

Comparing Japan's with China's learning coefficient, it is evident that though the value of Japan's coefficient is higher, it continued to decrease, suggesting that

[9] The authors discovered that the dynamic learning coefficient denotes the effect of market learning.

Table 2. Comparison of Dynamic Learning Coefficients of Japan's PC Technology (1982–2002)

	$\ln A$	a_0	a_1	a_2	a_3	a_4	d	$adj.\ R^2$	DW	AIC
$n=2$	1.91	0.080	–0.018	$2.2*10^{-4}$			0.27	0.923	0.56	–1.73
	(1.04)	(0.20)	(–1.08)	(0.52)			(1.04)			
$n=3$	7.66	–1.40	0.11	$-8.5*10^{-3}$	$2.1*10^{-4}$		0.68	0.979	1.38	–28.30
	(5.90)	(4.53)	(5.13)	(6.34)	(6.60)		(4.74)			
$n=4$	5.63	–0.84	0.039	$-6.8*10^{-5}$	$-2.5*10^{-4}$	$8.9*10^{-6}$	0.47	0.979	1.29	–27.81
	(2.35)	(1.33)	(0.52)	(0.008)	(–0.55)	(1.01)	(1.88)			

Data source: World Telecommunication Indicators Database 2004 (ITU) and World Development Indicators 2002

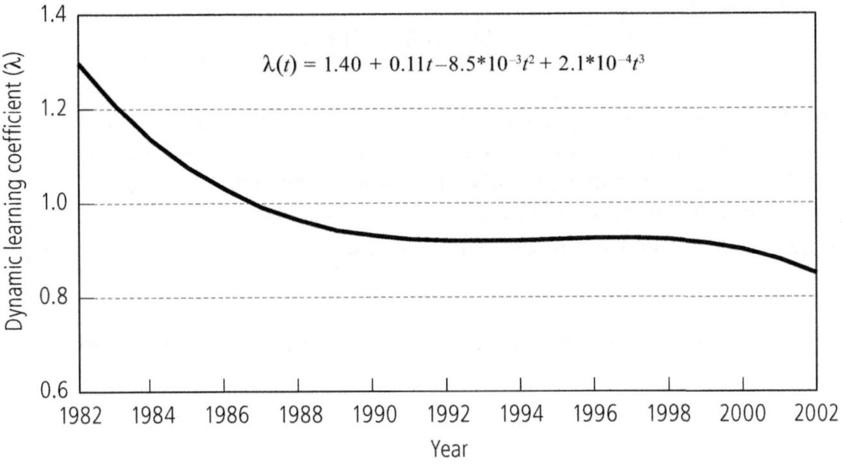

Fig. 12. Trend in the Dynamic Learning Coefficient in Japan's PC Technology (1982–2002)

Japan's PC industry has already shifted from the market learning to the functionality development phase.

3.3 *Price Change Due to Imported PC Parts and International Competition*

While market learning has played a significant role in price changes in China's PC industry—as demonstrated in Table 1 and indicated in the analytical framework of section 2—the impact on price changes of PCs also includes a disturbance term independent from market learning. This term consists of the effect of technology spillover ($\lambda(t)$) and price competition from international manufacturers ($D(t)$), depicted in equation (25).

$$\ln P = \ln A - \lambda(t) \ln PC + \delta(t) \ln IM + cD(t) \quad (25)$$

where A is scale factor; $\delta(t)$ is elasticity of imported parts to price; and c is the coefficient.

Given that the disturbance term is independent from market learning, equation (25) can be modified as follows:

$$\begin{aligned} \ln P + \lambda(t)\ln PC &= \ln A + \delta(t)\ln IM + cD(t) \\ &= \ln A + \sum_{i=0}^{n} b_i t^i \ln IM + cD(t) \\ &= \ln C + \sum_{i=0}^{n} b_i t^i \ln IM + c\frac{1}{1+e^{-at-b}} \end{aligned} \quad (26)$$

As demonstrated in the preceding section, $-b/a = 10, \delta(16) \geq 0$ in China's PC industry, using a heuristic approach, $n = 5$ for $\delta(t)$ and $a = 0.9$ (consequently, $b = -9.0$) for $D(t)$ are identified as statistically most significant.

After establishing these values, a regression analysis was conducted for the period 1982–2002. The result is summarized in Table 3.

Table 3. Elasticity of Imported Parts to Price and the Price Competition Function (1982–2002)

$$\ln P + \lambda(t)\ln PC = \ln A + \sum_{i=0}^{5} b_i t^i \ln IM + cD(t) + dD_{81}, D(t) = \frac{1}{1+e^{-0.9t+9.0}}$$

D_{81}: Dummy variable (1981 = 1, other years = 0).

a	$\ln A$	b_0	b_1	b_2	$b+3$	$b+4$	b_5	c	d	adj. R^2	DW	AIC
0.8	12.00	–0.45	0.38	–0.099	$9.6*10^{-3}$	$-4.0*10^{-4}$	$6.0*10^{-6}$	0.63	–0.13	0.971	2.71	–85.5
	(914.54)	(–7.98)	(6.04)	(–5.54)	(5.24)	(–5.02)	(4.85)	(4.43)	(–3.88)			
0.9	12.01	–45	0.38	–0.099	$9.6*10^{-3}$	$-4.0*10^{-4}$	$6.0*10^{-6}$	0.58	–0.13	0.973	2.72	–86.37
	(914.90)	(–8.17)	(6.22)	(–5.70)	(5.40)	(–5.17)	(4.99)	(4.59)	(–3.90)			
1.0	12.02	–0.45	0.39	–0.098	$9.4*10^{-3}$	$3.9*10^{-4}$	$5.8*10^{-6}$	0.52	–0.13	0.973	2.65	–86.32
	(894.40)	(–8.16)	(6.22)	(–5.69)	(5.38)	(5.15)	(4.95)	(4.58)	(3.86)			

Data source: World Telecommunication Indicators Database 2004 (ITU); World Development Indicators 2002; China Computer Newspaper and China Computer Education Newspaper

The effects of technology spillover from imported PC components on price are expressed by the elasticity function of imported parts to price ($\delta(t) = \frac{\partial \ln P}{\partial \ln M}$). Based on the results from Table 3, the trend in the effects of technology spillover from imported PC parts on China's PC technology from 1982 to 2002 is illustrated in Fig. 13.

Fig. 13 reveals that China fully benefited from technology spillover from imported PC components, particularly in the latter half of the 1980s, and peaking in the early 1990s. The positive effects of technology spillover subsequently began to diminish and ceased entirely after 1996. These observations demonstrate that 'the technological development of China's PC industry can be attributed to the effective

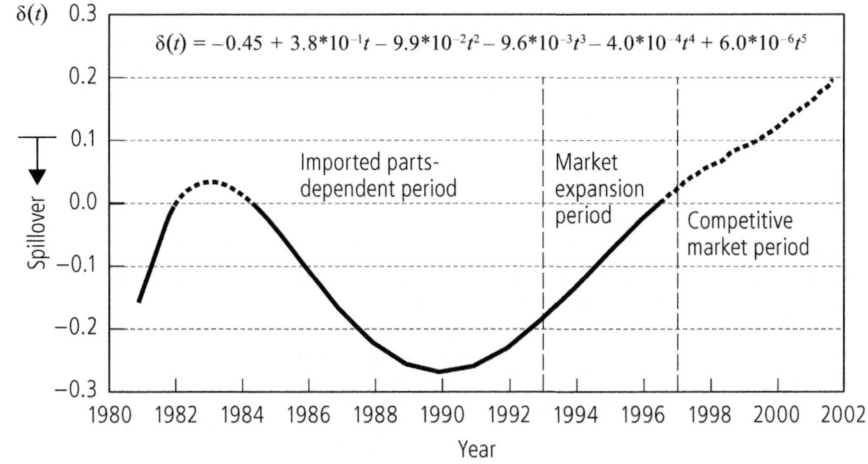

Fig. 13. Trends in the Effects of Technology Spillover from Imported PC Parts on China's PC Technology (1982–2002)

utilization of technology spillover by installing imported PC components into national products' during the imported parts-dependent period (1981–1992). Hence, hypothesis 1 can also be substantiated.

3.4 Trends in Creating Functionality of PC Innovation

3.4.1 Trends in Elasticity of PCs to GDP per Capita

As demonstrated by equations (13) and (14), GDP per capita in PC-developing countries can be described by the following equation:

$$\ln V = \ln B + \eta(t) \ln PC = \ln B + \sum_{i=0}^{n} c_i t^i \ln PC \tag{27}$$

Using regression analysis for the period 1981–2002, the coefficients with n=4 are identified as statistically most significant for the elasticity of PCs to GDP per capita in China. The resulting equation (28) is summarized in Table 4.

$$\eta(t) = 0.23 + 0.26t - 6.9 * 10^{-3}t^2 + 5.3 * 10^{-4}t^3 + 1.2 * 10^{-5}t^4 \tag{28}$$

Based on equation (28), the trend in elasticity of PCs to GDP per capita in China from 1981 to 2002 is illustrated in Fig. 14.

Fig. 14 reveals that elasticity peaked in 1992 (*at the end of the imported parts-dependent period*) and decreased in relation to the *shift from technology spillover to*

Table 4. Comparison of Elasticity of PCs to GDP per Capita in China (1981–2002)

	ln *B*	c_0	c_1	c_2	c_3	c_4	c_5	*b*	*adj.* R^2	DW	*AIC*
$n=3$	8.34	0.18	0.014	$-1.2*10^{-3}$	$2.8*10^{-5}$			–0.068	0.988	0.75	–63.60
	(417.37)	(6.10)	(1.00)	(0.90)	(0.78)			(0.76)			
$n=4$	8.36	0.23	0.026	$-6.9*10^{-3}$	$5.3*10^{-4}$	$1.2*10^{-5}$		–0.078	0.993	1.38	–75.04
	(524.39)	(12.87)	(1.93)	(2.55)	(2.87)	(3.01)		(1.94)			
$n=5$	8.34	0.29	–0.062	0.013	$-9.8*10^{-4}$	$2.5*10^{-5}$	$-1.3*10^{-7}$	0.13	0.991	1.18	–70.35
	(425.81)	(3.77)	(–1.09)	(0.93)	(–0.65)	(0.37)	(0.11)	(0.99)			

B is a coefficient for dummy variable D_91 (1991=1, other years=0)
Data source: World Telecommunication Indicators Database 2004(ITU); World Development Indicators 2002; China Computer Newspaper, and China Computer Education Newspaper

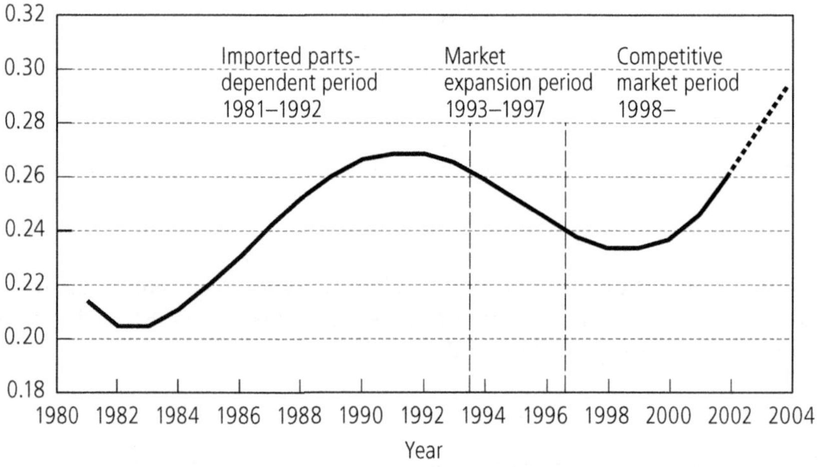

Fig. 14. Trends in Elasticity of PCs to GDP per Capita in China (1981–2002)

market learning. The elasticity of PCs began to increase again from 1998 onward, thus corresponding to the onset of *the competitive market period*.

Fig. 15 demonstrates a high correlation between elasticity of PCs to GDP and PC functionality (MPU). High functionality is represented by the speed of the basic function of PCs, namely, MPUs. In 2000, China's PC manufacturers installed MPUs of 2.4GHz, up from 0.35GHz in 1998, which implies that China's PC makers are making an effort to innovate the functionality of their PCs.

3.4.2 International Comparison

To verify that China's PC industry has progressed to the high functionality development phase, the elasticity of PCs to GDP per capita among 15 PC-developing countries from 1981 to 2002 was established using equation (27). Table 5 summarizes the results of the analysis.

Table 5. Measurement of Elasticity of PCs to GDP per Capita in 15 PC-Developing Countries (1981–2002)

		ln*B*	c_0	c_1	c_2	c_3	c_4	c_5	*b*	*adj.* R^2	*DW*	η in 2002
China	$n = 4$	8.36	0.23	0.026	$-6.9*10^{-3}$	$5.3*10^{-4}$	$1.2*10^{-5}$		–0.078	0.993	1.38	0.252
		(524.39)	(12.87)	(1.93)	(2.55)	(2.87)	(3.01)		(1.94)			
Australia	$n = 4$	8.39	0.25	0.02	$-2.8*10^{-3}$	$1.6*10^{-4}$	$-3.1*10^{-6}$		0.04	0.983	1.31	0.234
		(24.0	(2.99)	(2.84)	(3.13)	(3.17)	(3.10)		(1.27)			
South Korea	$n = 5$	8.04	–0.14	0.13	–0.02	$1.5*10^{-3}$	$-5.9*10^{-5}$	$8.5*10^{-7}$		0.993	2.14	0.169
		(170.54)	(0.78)	(1.80)	(–1.65)	(1.54)	(1.46)	(1.39)				
India	$n = 5$	5.91	0.12	0.017	$-6.4*10^{-3}$	$7.8*10^{-4}$	$-3.7*10^{-5}$	$6.2*10^{-7}$		0.997	2.58	0.161
		(122.2)	(11.36)	(1.77)	(2.24)	(2.33)	(2.22)	(2.06)				
Spain	$n = 5$	8.93	0.11	–0.02	$8.1*10^{-3}$	$-9.0*10^{-4}$	$4.3*10^{-5}$	$-7.1*10^{-7}$		0.997	1.58	0.126
		(105.72)	(1.96)	(–3.37)	(3.61)	(–3.87)	(4.11)	(–4.27)				
Malaysia	$n = 5$	7.56	–0.064	0.082	–0.018	$2.0*10^{-3}$	$-8.6*10^{-5}$	$1.4*10^{-6}$		0.984	1.74	0.109
		(113.42)	(–0.61)	(1.98)	(–2.07)	(2.31)	(–2.48)	(2.58)				
Indonesia	n = 4	6.34	0.214	–0.040	$8.8*10^{-3}$	$6.3*10^{-4}$	$1.34*10^{-5}$			0.978	1.41	0.107
		(430.02)	(6.56)	(1.77)	(2.28)	(2.63)	(2.80)					
Thailand	$n = 5$	6.90	0.21	0.09	–0.03	$2.8*10^{-3}$	$-1.3*10^{-4}$	$2.1*10^{-6}$		0.990	1.48	0.107
		(430.08)	(1.75)	(1.01)	(–1.29)	(1.51)	(–1.69)	(1.84)				
South Africa	$n = 5$	8.25	–0.26	0.09	–0.02	$1.5*10^{-3}$	$-5.8*10^{-5}$	$8.1*10^{-7}$	–0.03	0.945	1.34	0.119
		(213.66)	(3.01)	(2.46)	(2.52)	(2.44)	(2.27)	(2.08)	(1.78)			
Mexico	$n = 5$	8.27	0.16	–0.028	0.010	$-1.3*10^{-3}$	$6.2*10^{-5}$	$-1.1*10^{-6}$		0.902	1.95	0.110
		(146.32)	(5.72)	(–2.21)	(2.95)	(3.22)	(3.33)	(3.36)				
Russia	$n = 5$	8.05	–0.19	–0.04	0.02	$-2.4*10^{-3}$	$1.3*10^{-4}$	$-2.4*10^{-6}$	0.03	0.937	2.43	0.101
		(90.86)	(3.07)	(1.04)	(1.41)	(1.71)	(1.97)	(2.19)	(0.49)			
Turkey	$n = 5$	7.59	–0.062	0.14	–0.029	$2.6*10^{-3}$	$-1.0*10^{-4}$	$1.6*10^{-6}$	–0.066	0.990	2.97	0.112
		(82.07)	(–0.55)	(2.37)	(–2.70)	(2.86)	(2.92)	(2.90)	(–6.12)			
Greece	n = 5	8.82	0.098	–0.018	$6.3*10^{-3}$	$-7.5*10^{-4}$	$3.7*10^{-5}$	$-6.3*10^{-7}$		0.993	1.94	0.104
		(100.85)	(2.12)	(–3.03)	(4.08)	(4.84)	(5.35)	(5.63)				
Colombia	$n = 5$	7.21	0.12	0.023	$-7.3*10^{-3}$	$8.9*10^{-4}$	$-4.6*10^{-5}$	$8.3*10^{-7}$		0.989	2.44	0.099
		(197.60)	(3.31)	(5.95)	(4.32)	(2.50)	(2.88)	(3.15)				
Philippines	$n = 4$	6.74	0.081	0.012	$-3.3*10^{-3}$	$2.3*10^{-4}$	$-4.6*10^{-6}$		0.042	0.970	1.64	0.110
		(109.8)	(8.65)	(1.90)	(3.38)	(3.89)	(3.97)		(6.03)			

Data source: World Telecommunication Indicators Database 2004; World Development Indicators 2002

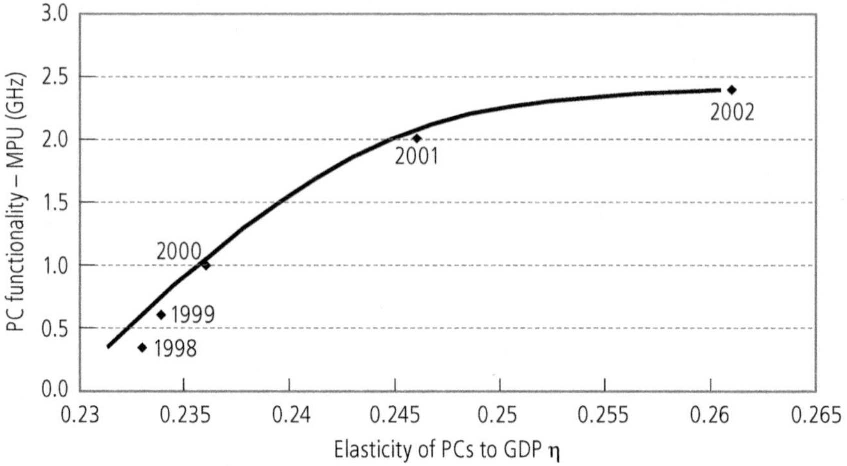

Fig. 15. Correlation between Elasticity of PCs to GDP per Capita and PC Functionality in China (1981–2002)

Table 5 denotes that the elasticity of PCs to GDP per capita is highest in China among the 15 PC-developing countries examined, followed by Australia, South Korea, India, and Spain.

These analyses, depicted in Figs. 14 and 15, as well as in Table 5, substantiate hypothesis 3 that 'confronted with the competitive market, China's PC industry shifted from technological advancement through market learning to high functionality development, with the objective of catching up with international competitors.'

4 Conclusions

This chapter has aimed to identify the institutional reasons for the initial inertia characterizing China's PC industry, which has evolved along a continuously rising trajectory and is now extremely successful, both in terms of quantity and quality.

A thorough review of China's institutional system consisting of the national strategy, the socio-economic system, entrepreneurial organization, culture, and historical perspectives, exposed the following valuable insights:

1. The development trajectory of China's PC industry can be divided into three phases:

    a. Imported parts-dependent period (1981–1992),
    b. Market expansion period (1993–1997), and
    c. Competitive market period (1998–present).

2. PC components exported by international manufacturers dominated the first phase of development, the *imported parts-dependent period.*
3. During this first period, technology spillover from imported parts was incorporated, thus enhancing the technological level of China's PCs.
4. As the diffusion of PCs increased, imported PC components were increasingly substituted by indigenous ones, leading to the second phase, the *market expansion period.*
5. In this period, market learning as a consequence of cumulative learning in the course of production, distribution, and utilization functioned effectively to further enhance the technological level of PCs produced in China, resulting in the third phase, the *competitive market period.*
6. Facing severe competition from international PC makers in the competitive market, indigenous PC manufacturers focused on creating high functionality with the objective of catching up with international products.

These observations led to the assumption that the initial inertia of China's PC industry generated the upward trajectory, leading to its current success:

1. Triggered by the effective utilization of technology spillover from imported PC components;
2. Followed by the acceleration of market learning effects; and
3. Shifting to high functionality development in light of market competition, thereby catching up with international competitors at the end of the 1990s.

We corroborated our hypothetical assumptions by empirically analyzing the development trajectory of China's PC industry in the last two decades, focusing on China's institutional system.

Given the distinct features of China's institutional system that drive the country's high dependency on PCs and, subsequently, the remarkable progress the PC industry has achieved within its phased development, the role of this distinct system with a particular focus on the following dimensions should be examined more closely in future studies:

1. The multiplier effects of a phased development (technology spillover, market learning, and high functionality development), activating the development of the PC industry, and the diffusion that induced the enhancement of innovative PC functionality, and
2. Coevolution between this activated development in the PC sector and China's distinct institutional structure, which accelerated the enhancement of innovative PC functionality.

References

Abernathy W J (1978) The productivity dilemma: Roadblock to innovation in the automobile industry. Johns Hopkins University Press, Baltimore

Antonelli C (2006) Localized technological change and factor markets: Constraints and inducements to innovation. Structural Change and Economic Dynamics 17:224–247

Cantner U, Pyka A (2001) Classifying technology policy from an evolutionary perspective. Research Policy 30:759–775

Checkland P B (1985) From optimizing to learning a development of systems thinking from the 1990s. Journal of Optimal Research Society 9:757–767

Checkland P B (1988) Images of systems and systems image. Journal of Applied Systems Analysis 15:37–42

Coe D T, Helpman E (1995) International R&D spillovers. European Economic Review 39:859–887

Denison E F (1962) The sources of economic growth in the U.S. and the alternatives before U.S. Committee for Economic Development, New York

Gatignon H, Robertson T S (1991) Innovative decision process. In: Robertson T S, Kassarjan H H (eds) Handbook of consumer behavior, Prentice Hall, Englewood Cliffs

Gomulka S (1971) Inventive activity, diffusion, and the stages of economic growth. Institute of Economics, Aarhus University, Aarhus

Gregory F N (2006) Beyond the learning curve: Factors influencing cost reductions in photovoltaics. Energy Policy 34:3218–3232

Griliches Z (1957) Hybrid corn: An exploration in the economics of technological change. Econometrica 25(4): 501–522

Griliches Z (1979) Issues in assessing the contribution of R&D to productivity growth. Bell Journal of Economics 10:92–116

Holmes T J, Schmitz J A (2001) Competition at work: Railroads vs. monopoly in the U.S. shipping industry. Quarterly Review 25(2):3–29

Janes F R (1995) Interactive management: Framework, practice and complexity. In: Ellis et al. (eds) Critical issues in systems theory and practice, Plenum Press, New York

Jorgenson D W, Griliches Z (1967) The explanation of productivity change. Review of Economic Studies 34(3):249–283

Jovanovic B, Lach S (1989) Entry, exit, and diffusion with learning by doing. The American Economic Review 79(4):690–699

Katz M L, Shapiro C (1985) Network externalities, competition, and compatibility. American Economic Review 75(3):424–440

Keller W (1996) Absorptive capacity: On the creation and acquisition of technology in development. Journal of Development Economics 49:199–227

Lai M, Peng S, Bao Q (2006) Technology spillovers, absorptive capacity and economic growth. China Economic Review 17:300–320

Lumenga-Neso O, Olarreaga M, Schiff M (2005) On 'indirect' trade-related R&D spillovers. European Economic Review 49(7):1785–1798

OECD (1997) Technology and industrial performance. OECD, Paris

OECD (1998) Technology, productivity and job creation: Best policy practices. OECD, Paris

Oren S S, Smith S A (1981) Critical mass and tariff structure in electronic communications markets. Bell Journal of Economics 12(2):467–487

Rogers E M (1983) The diffusion of innovations, 3rd edn. The Free Press, New York

Venkatraman M P (1988) Investigating differences in the roles of enduring and instrumentally involved consumers in the process. Advances in Consumer Research 15:299–303

Verhulst P F (1845) Recherches mathematiques sur la loi d'accroissement de la population. Nouveaux memories de l'Academie Royale des Sciences, des Lettres et des Beaus-Arts de Belgique 18:1–32

Watanabe C (1999) Systems option for sustainable development. Research Policy 28(7): 719–749

Watanabe C (2005) Institutional MOT: Co-evolutionary dynamism of innovation and institution. Proceedings of IAMOT, Vienna, 2005

Watanabe C, Asgari B, Nagamatsu A (2003) Virtuous cycle between R&D, functionality development and assimilation capacity for competitive strategy in Japan's hi-technology industry. Technovation 23(11):879–900

Watanabe C, Matsumoto K, Hur J Y (2004) Technological diversification and assimilation of spillover technology: Canon's scenario for sustainable growth. Technological Forecasting and Social Change 71(9):941–959

Watanabe C, Asgari B (2004) Impacts of functionality development on the dynamism between learning and diffusion of technology. Technovation 24:651–664

An Empirical Analysis of the Coevolution of China's Institutional System and Rapidly Growing PC Sector

Chihiro Watanabe, Shanyu Lei, Akihisa Yamada, and Masanobu Kita

Abstract Since its establishment 20 years ago, China's personal computer (PC) industry has undergone a significant development by leveraging technology spillover through imports of PC parts and accelerated market learning. Despite its extraordinary development, China's PC industry is now facing strong competition from foreign producers, who are responding to nonlinear progress by attempting to establish high functionality development. Functionality development is generally defined as the ability to considerably improve the performance of production processes, goods, and services by means of innovation (Watanabe et al., 2005). Furthermore, the particular attributes of China's social and economic institutions have an impact on the education of the population, the level of urbanization, the national culture's degree of openness, the entrepreneurial culture, etc. These observations suggest the following hypotheses on China's institutional system, which has taken China's PC development to a new level: (i) The multiplier effects of a phased development (technology spillover, market learning, and high functionality development) have contributed to the rapid increase in the development of PC technology and diffusion, thus generating an improvement in the functionality of PC innovation; and (ii) The coevolution of rapid increase in the development of PC technology and China's distinct institutional system has accelerated the improvement of the functionality of PC innovation. (iii) Based on an empirical analysis of the development and diffusion trajectory of China's PC industry focusing on the impact of various distinct institutional characteristics, this chapter attempts to verify these hypotheses.

Corresponding Author: Chihiro Watanabe
Department of Industrial Engineering & Management, Tokyo Institute of Technology, 2-12-1, W9-49, Ookayama, Meguro-ku, Tokyo 152-8552, Japan, and International Institute for Applied Systems Analysis (IIASA), Laxenburg, Austria, e-mail: chihiro@me.titech.ac.jp

Shanyu Lei, Akihisa Yamada and Masanobu Kita
Department of Industrial Engineering and Management, Tokyo Institute of Technology, Tokyo, Japan

M. V. Van Geenhuizen et al., (eds). *Technological Innovation Across Nations*,
DOI: 10.1007/978-3-642-00158-1_2,

Key words: Institutional Systems, Technology Spillover, Market Learning, China's PC Development

1 Introduction

China's PC industry has undergone a striking development in the last 20 years. This development trajectory can be divided into three phases: i) imported PC parts-dependent period (1981–1992); ii) market expansion period (1993–1997); and iii) competitive market period (1998–present). The three phases are illustrated in Fig. 1.

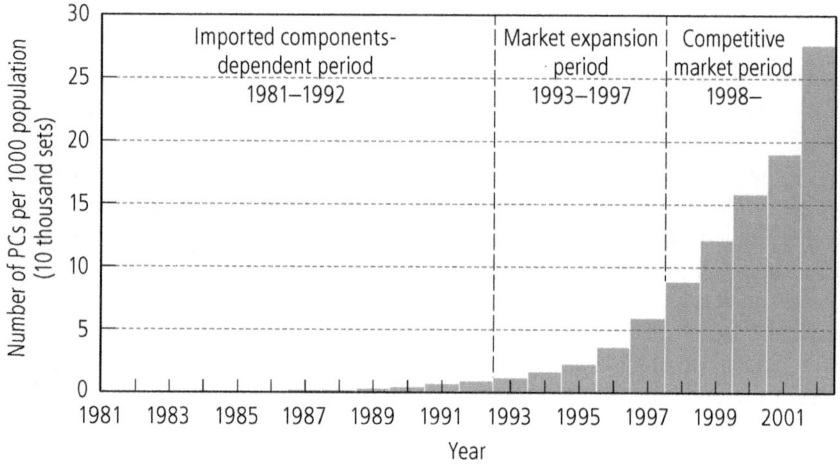

Fig. 1. Number of PCs per 1000 Population (1981–2002)
Source: World Telecommunication Indicators Database 2004, International Telecommunication Union

Analysis by Lei et al. (2009) reveals that spillover technology from imported PC parts significantly enhanced the technological level of China's PC industry during the first period.

PC production increased dramatically during the second period, inducing market learning as a consequence of cumulative learning in the course of PC production, diffusion, and utilization. These factors functioned effectively to further increase the technological level of China's PC industry and resulted in the third phase, the competitive market period.

China's PC industry has shifted from technological advancement through market learning to the development of innovations, leading to a considerable improvement in the sector's performance, thus 'catching up' with foreign competitors.

As a result of this phased progression, China's PC development[1] resulted in the country reaching the world's highest average annual growth rate of 42.9% in terms of PC diffusion.

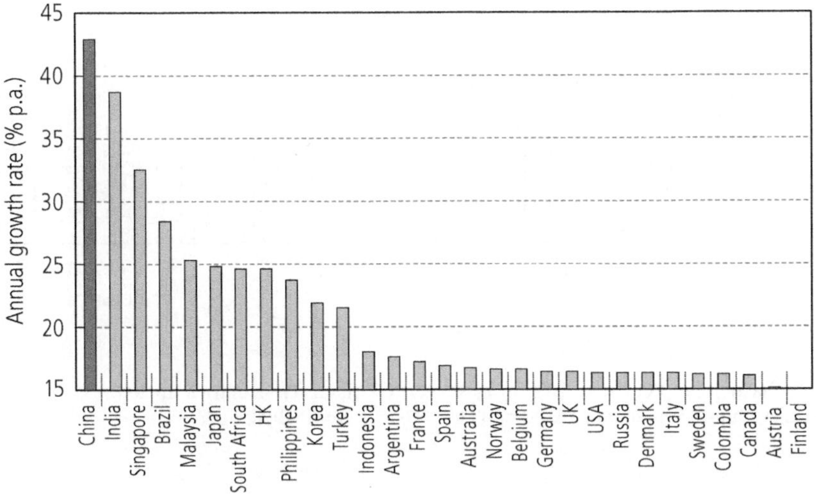

Fig. 2. International Comparison of Growth in Number of PCs (average 1993–2001)
Source: World Information Technology and Services Alliance (2003)

This remarkable achievement can be attributed to the highly conducive environment for PC development, aiming to considerably increase PC performance as demonstrated in Fig. 3.

Fig. 3 compares the elasticity of PC use to GDP (i.e., the relation between change in GDP per capita and changes in the number of PC users per 1000 inhabitants) in 15 countries with a low PC penetration. The degree to which countries with a low diffusion of information communication technology (ICT) focus on increasing PC use and catching up with foreign competitors by improving the functionality of indigenous products, as well as the degree of elasticity of PC use in relation to GDP, sheds light on the extent of efforts undertaken to advance functionality development. Fig. 3 shows that China's level of elasticity is the highest among countries with a low degree of PC penetration, indicating that China offers a highly conducive environment for PC development and the improvement of PCs' performance.

These observations suggest the following hypotheses about distinct features of China's institutional system that have led its PC development to surge to a new level:

[1] "PC development" refers here to the development of China's PC industry and the diffusion of PCs.

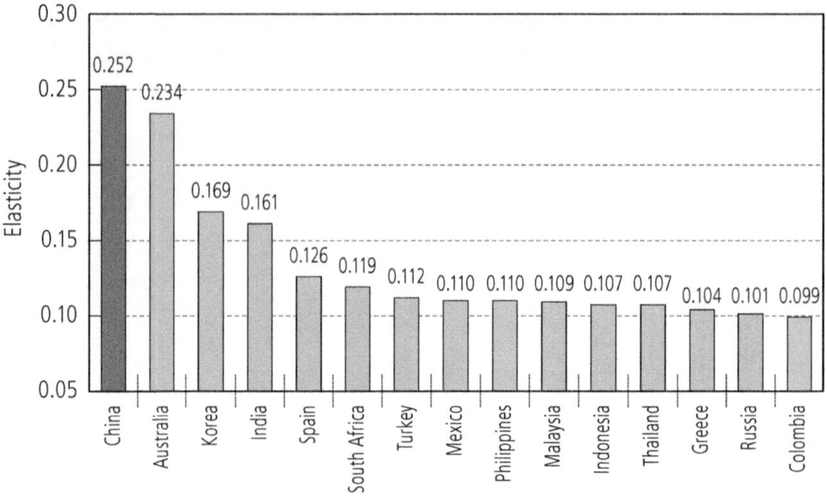

Fig. 3. Elasticity of PC Use to GDP in 15 Lower ICT Intensity Countries (in 2002)
Source: Authors' computation based on ITU, UN, and World Bank figures

1. The multiplier effects of a phased development (technology spillover, market learning, and high functionality development) have rapidly increased PC development and diffusion, thus inducing an enhanced functionality with respect to PC innovation, and
2. The coevolution of a rapidly increasing PC development and China's particular institutional system has accelerated this enhanced functionality of PC innovation.

To date, many studies have attempted to identify the dynamics that induce a surge in new innovations. Innovation, assimilation, and utilization of technology largely depend on a set of interacting factors including economy, society, culture, customs, educational system, and policy. This comprehensive system is defined as the institutional environment. North (1994, p. 360) postulated that the "humanly devised constraints that structure human interaction [...] are made up of formal constraints (e.g. rules, laws, and constitutions), informal constraints (e.g. norms of behavior, conventions, and self-imposed codes of conduct) and their enforcement characteristics."

In this chapter, the social and economic impacts of institutions are classified according to three dimensions: (i) National strategy and socio-economic system, (ii) Entrepreneurial organization and culture, and (iii) Historical perspectives (Watanabe, 2005). The influence of institutions on a country's ability to master and advance technology fundamentally affects economic performance (Nelson and Nelson, 2002). National strategy, including the government's policy, also influences economic performance considerably (Yang et al., 2006).

Watanabe and Kondo (2003) demonstrated that the nature of IT could be developed in a self-propagating way through its interaction with institutions. Janes

(1995) and Flood (1987) contend that while complex institutions reject the concept of coevolution with innovation, it may accelerate a self-propagating development once it passes a certain threshold. Flood (1987) also pointed out that the complexity of institutions may consist of system-driven complexity, as well as the indigenous complexity of the given nation. UNESCO (1998) asserted that the complexity of institutions is indigenous and changes with the nation's mindset, preferences, and values. In this context, the attainment of higher education is considered to significantly influence both the national strategy and the socio-economic system, as evidenced in Japan and South Korea, where substantial investment in education has been critical for achieving high economic performance (Lee, 2003; Douangngeune et al., 2005).

Oren and Smith (1981) suggest that the self-propagation of IT can be attributed to network externalities and the effects of interaction between supply and demand in the market. That is, if a complex institution with a high degree of interaction between supply and demand exceeds the critical mass, the effects of network externalities may rapidly evolve. Similarly, Janes (1995) has claimed that the stimulation of optimal interaction (i.e., the interaction between supply and demand through market learning and indigenous development) may accelerate the establishment of complex institutions, which in turn results in the exceedance of critical mass. Flood (1987) postulated that institutions' complexity rests on a system of different dimensions of complexity determined by the institutional system and the intricate nature of people's abilities, interests, standards, feelings, sense of value, and beliefs. In short, previous research suggests that complex institutions possibly obstruct a coevolution with innovation, but may, on the other hand, accelerate self-propagation once externality effects lead to the exceedance of critical mass. While complexity is an institution-specific attribute, research also confirms that the complexity can be modified in accordance with changes in perception and the sense of value of both firms and customers.

During the 1960s, when extensive studies were carried out to identify the incentives behind innovation, the concept of "induced bias" (adjustments of discrepancies in the supply conditions) was developed (Kennedy, 1964; Weizsaecker, 1966; Phelps, 1966). The authors asserted that the incentive to correct an induced bias (primarily the gap between production factors) derived from a strong inducement of technological innovation. In line with this assumption, the primary incentive for further development within a nation is linked to the country's development stage. While countries that have already progressed to a mature economy focus on improving social satisfaction, countries that have not yet reached this goal are preoccupied with decreasing the bias (i.e., closing the gap between production factors) at home and catching up with more developed countries to reduce the existing bias between them. These incentives are strongly linked to institutional complexity and technological innovation.

According to Teece (1986), indirect support services of innovations such as competitive production methods, distribution channels, marketing, and after-sales support were indispensable in the development and diffusion of the PC industry. He defines such services as complementary assets. Such assets could be established within institutions in each respective country. If induced bias is considered the primary incentive for innovation, institutional complexity could—with reference to technological innovation—foster the establishment of complementary assets, which are key factors in the development and diffusion of PCs.

While the above-mentioned studies provide insight into our hypothesis on the distinct role China's social and economic institutions have played in stimulating the PC industry's rapid growth, they do not elucidate the complexity of China's distinct institutional system adequately. Therefore, an empirical analysis is carried out to shed light on the coevolutionary dynamics between China's distinct institutional structure and the development of its PC industry.

Section 2 reviews some of the characteristics of China's institutional structure and their socio-economic impacts. Section 3 presents the analytical framework, and the empirical analysis follows in Section 4. Finally, Section 5 summarizes our findings and their policy implications, as well as recommendations for future research.

2 China's Distinct Institutional Characteristics

Historically, China's institutional system has rested on a distinct set of causal relations between urbanization, higher income, and higher education, as illustrated in Fig. 4.

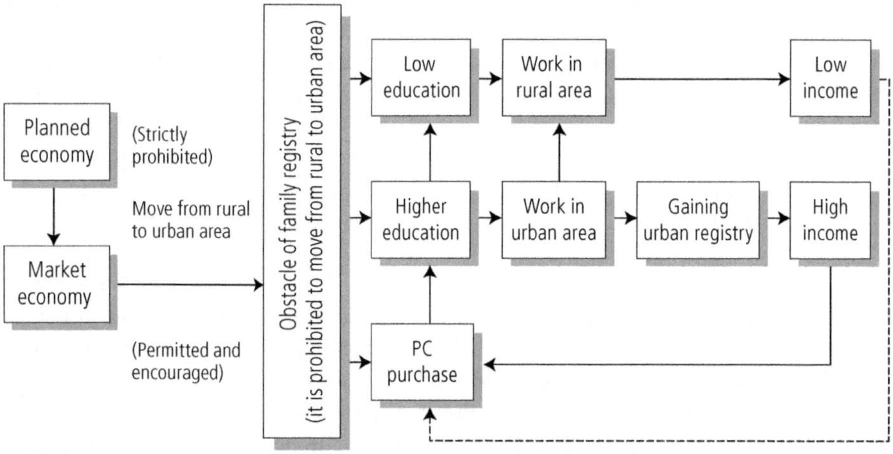

Fig. 4. Causal Relations between Urbanization, Higher Income, Higher Education, and Dependency on PCs in China

During the era of planned economy, moving from a rural to an urban area was strictly prohibited. The only way for an individual residing in a rural area to obtain an urban household register was to pursue a higher education. Students attending a university in an urban area received new residency permits and were entered into the urban household register, thus enabling students to seek employment in an urban area after graduating. Although this barrier has been largely eradicated and moving from rural to urban areas has in principle been encouraged with the emergence of the market economy, certain institutional barriers preventing rural residents from moving to urban areas continue to remain. The primary purpose of purchasing a PC under such circumstances is to expand children's knowledge and skills to increase their odds of attaining a higher education. In most cases, PCs are purchased by families residing in rural areas with a low household income, making the PC a very expensive investment. Despite the financial burden, parents buy PCs to enhance their children's knowledge and increase their likelihood of passing the required entrance examinations to institutes of higher learning, thus making the acquisition of an urban residency permit and working and living in an urban area possible in the future. The reasons for purchasing a PC in China differ from those in developed countries, where PCs are a means to improve lifestyle and not necessarily to advance children's education. Consequently, the dependency on PCs generates China's distinct institutional structure with regard to ICT development. It can thus be concluded that the coevolutionary relationship between this distinctive institutional structure and activated PC development plays a significant role in accelerating the enhancement of quality and performance of PCs in China.

Table 1 compares major socio-economic indicators underlying institutional systems—as defined in this chapter—that relate to ICT in different countries. Nations' ICT development is primarily influenced by these institutional systems and their impacts. Table 1 suggests that China's institutional system is quite particular with regard to ICT development.

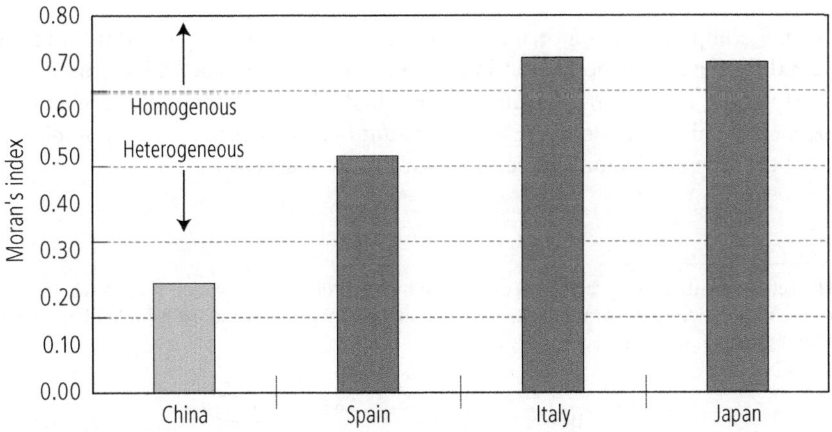

Fig. 5. Regional Differences by Moran Index (2000)
Source: Caniels (2000)

Table 1. Socio-Economic Indicators Relating to ICT

	China	Japan	India	USA	Europe
Population (in 10,000)[a]	128,453	12,692	101,592	28,142	8,215 (Germany[1])
Proportion of people living in urban areas (%)[a]	37.7	65.3	28.1	79.8	87.9 (Germany)
Regional differences[a]	high	low	–	–	Low: Italy Med.: Spain
Inequality of income distribution[a]	40.3	24.9	37.8	40.8	38.2 (Germany)
Attainment of higher education[b]	5.0	47.7	8.0	39.1	21.8 (Germany)
Education children acquire at home (on their own: full score is 100)[b]	86	65	30	53	73 (Germany)
Number of users of Internet (per 1000 inhabitants)[c]	46.0	448.9	15.9	551.4	411.9 (Germany)
Degree of nations' ICT utilization (Information and communication technologies are an overall priority for the government: 1= strongly disagree; 7= strongly agree)[c]	1.37	4.12	1.17	4.63	4.75 (Germany)

Sources:
[a]Human Development Report (2004);
[b]World Competitiveness Yearbook (IMD 2004);
[c]The Global Information Technology Report (World Economic Forum 2004).
[1] Germany ranked third in the world in terms of total number of PCs installed in 2002 and its GDP is larger than any other European country's; therefore, Germany was selected as an example for Europe.

Fig. 5 compares the regional differences between China, Spain, Italy, and Japan in 2000 by means of the Moran index.[2] Fig. 5 indicates that China's index is the lowest among the countries examined, and that China has the highest level of heterogeneity. This suggests that China's institutional structure is more complex than that of the advanced countries selected for the comparison.

[2] The heterogeneity among regions is calculated by the Moran Index, which reveals whether a spatial auto-correlation exists or not. A lower value in the Moran Index indicates higher heterogeneity among regions. It is computed as follows:

$$I = \left(\frac{n}{W}\right)\frac{\sum_{i=1}^{n}\sum_{j=1}^{n}(W_{ij}z_iz_j)}{\sum_{i=1}^{n}z_i^2}, \text{and } W = \sum_{i=1}^{n}\sum_{i=1}^{n} w_{ij}$$

where w_{ij}: the weight of regions, $w_{ij}=1$: similarity among regions and $w_{ij}=0$: difference among regions.

The reason China, Spain, Italy, and Japan are compared is to examine the complex institutional structure due to the heterogeneity among regions in China as compared to other advanced countries.

3 Analytical Framework

3.1 Changes in the Pace of PC Development and Diffusion

The emergence of a new functionality development in China's PC industry was triggered through the assimilation of spillover technology from imported PC components and competition with international manufacturers, thus maximizing the marginal productivity of technology $\frac{\partial PC_T}{\partial T}$. Where PC technology stock T ($\Delta T \approx R$) is created by accumulating R&D investment (R), an increase in technology stock enhances PC's functionality development, as demonstrated by the logistic growth curve in Fig. 6, which was derived from the epidemic function.

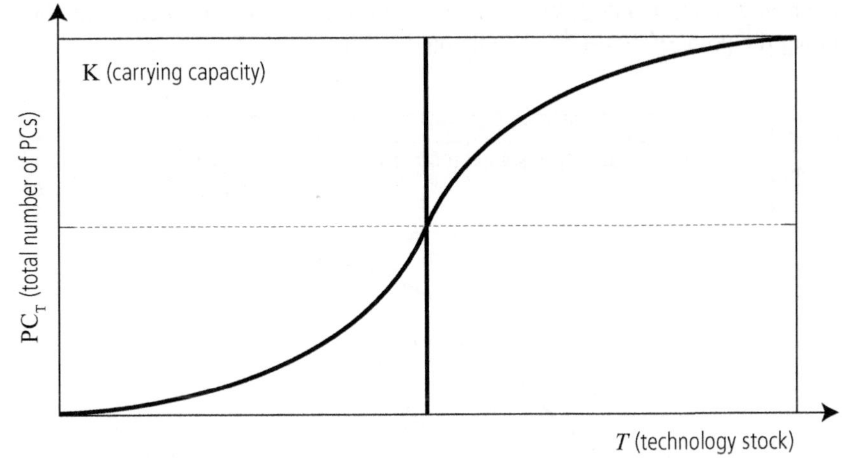

Fig. 6. Development Trajectory of PCs Enhanced by Technology Stock

The epidemic function in Fig. 6 is depicted by equation (1).

$$\frac{\partial PC_T}{\partial T} = aPC_T(1 - \frac{PC_T}{K}) \tag{1}$$

where a: velocity of diffusion; and K: carrying capacity: the invention and diffusion of technologies which raise the limit (Meyer and Ausbel, 1999).

The left-hand side of equation (1) indicates the marginal productivity of technology (MPT); its rate of change is described by equation (2).

$$\frac{\Delta MPT}{MPT} = aR(1 - \frac{2}{K/PC_T}) \tag{2}$$

As demonstrated in equation (1), a PC's development trajectory can be represented by the following logistic growth function:

$$PC_T = \frac{K}{1+e^{-aT-b}} \tag{3}$$

where b: coefficient, denoting the initial state of diffusion.

The state of marginal productivity of technology can be determined as follows, depending on the technology stock's position on the trajectory:

$$T < -\frac{b}{a} : \frac{\Delta MPT}{MPT} > 0;\ T = -\frac{b}{a} : \frac{\Delta MPT}{MPT} = 0;\ \text{and}\ T > -\frac{b}{a} : \frac{\Delta MPT}{MPT} < 0.$$

By being linked to the same development trajectory, PC development enters a vicious cycle—as marginal productivity decreases, the rate of PC increase declines and stagnation of technology sets in once the inflection point of $-\frac{b}{a}$ has been reached.

To avoid getting caught up in this vicious cycle, PC developers should aim to initiate a new development trajectory involving PCs with a higher functionality before reaching the point of $T > -\frac{b}{a}$, as illustrated in Fig. 7.

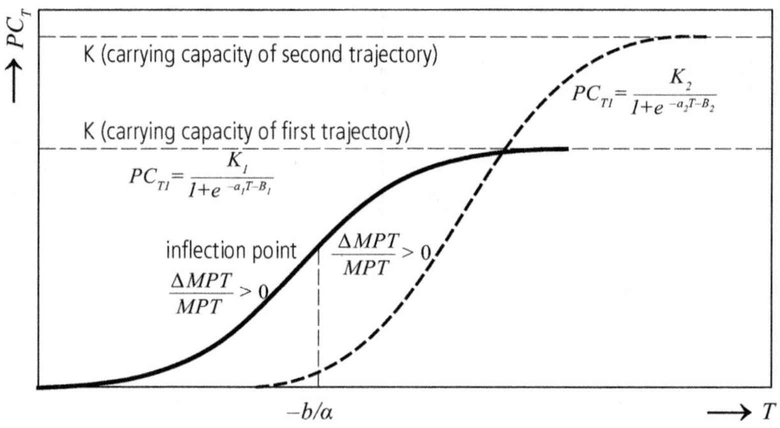

Fig. 7. Formulating a PC Development Trajectory

Accordingly, provided that the PC development trajectory can be traced by the foregoing logistic growth function, PC development would have the highest activation at inflection point $T > -\frac{b}{a}$, when the preceding PC development trajectory reaches the maturity stage, and a new trajectory emerges to complement the previous stagnant one (Watanabe et al., 2005).

Since PC technology stock increases over time in developing countries,[3] the development stage of PCs—based on the foregoing assumption with regard to the development trajectory of PCs—can be determined by assessing the inflection point $-b/a$ in relation to its time trend dimension t. Hence, since PC development in developing countries had not yet reached the inflection point in 2002, it can be defined as $-b/a > 2002$. The degree of activation of PC development π in 2002 increased as the distance between the inflection point ($t_{-b/a}$ when $t = -b/a$) and 2002 decreased. That is, π can be depicted by reversing the balance:

$$\pi = \frac{1}{t_{-b/a} - 2002} \tag{4}$$

3.2 Institutional System Characteristics Determining PC Development and Diffusion

As stated in Section 1, institutions are defined as three-dimensional systems, consisting of (i) national strategy and the social system, (ii) entrepreneurial organization and culture, and (iii) historical perspectives.

Given China's distinct institutional structure, which underlies the country's strong dependence on PCs, the motivation to move from rural to urban areas, and the reliance on PCs to access higher education (as discussed in Section 2), the following indices (prime factors) were categorized in accordance with each respective dimension: (i) *attainment of higher education*, nationwide *ICT use*, *number of Internet users*, and *ICT use in the public sector* for the first dimension; (ii) *enabling environment for ICT*, and *ICT use in the private sector* for the second dimension; and (iii) *population*, *ratio of people living in urban areas, cultural elasticity and adaptability to change, national culture (openness of the culture)*, as well as *political instability and social violence* for the third dimension. The relation between prime factors and the three dimensions of the PC industry's institutional system is illustrated in Fig. 8. The prime factors characterizing the institutional structure as depicted in Table 1 are found on the right of Fig. 8, while the indices of the three dimensions of the PC industry's institutional system that underlie the prime factors are on the left.

By evaluating the 11 indices representing the three dimensions of the PC industry's institutional system, through principal component analysis, the institutional structure relevant for PC development can be measured.

[3] PC developing countries refers to countries in which the early stage of PC development has not yet reached the inflection point on the PC development trajectory.

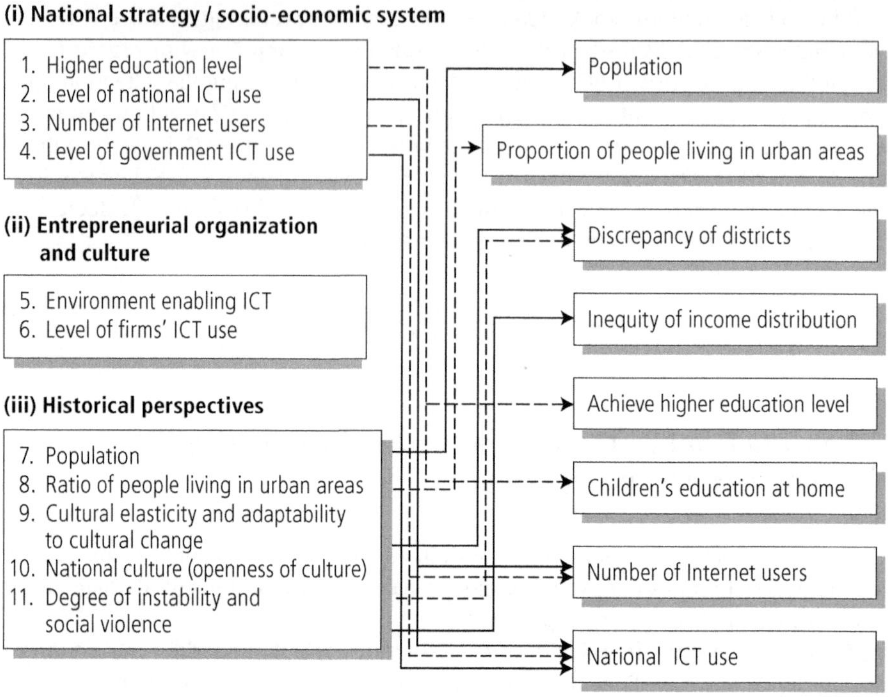

Fig. 8. Relationship between Prime Factors and the Dimensions of the PC Industry's Institutional System

4 Empirical Analysis

4.1 Multiplier Effects of Phased Development for the Enhancement of PC Development and Diffusion

4.1.1 Comparison of PC Development Stage and Enhancement of PC Development

A country's PC development and diffusion trajectory can be traced by a logistic growth curve derived from the epidemic function. Thus, PC development trajectories in 24 countries [4] from 1981 to 2002 were analyzed, each demonstrating statistical significance.[5] The stage of PC development can be examined by comparing their inflection points on the logistic growth curve. Fig. 9 compares the PC development

[4] Due to data constraints, the 24 countries selected included some OECD countries: Australia, USA, Canada, Turkey, Sweden, UK, Spain, Mexico, South Korea, Germany, Italy, France, Greece, and Japan; some developing countries: India, China, Indonesia, South Africa, Colombia, Philippines, Thailand, and Malaysia; and two other countries, Singapore and Russia.

[5] See the Appendix for details on the results of the analysis.

stage of 24 countries, indicating the respective year of each country's inflection point on a logistic growth curve. Fig. 9 suggests the following implications with respect to the activation of PC development in 2002:

a. China's PC development trajectory was estimated to reach the inflection point by January 2004, and its development stage in 2002 was closer to the inflection point than all the other countries that had not yet reached that point.
b. Thus, the growth rate of China's PC development was highest among the 24 countries examined, followed by Australia (inflection point in September 2004), South Korea (February 2005), and India (April 2007).
c. Japan and USA reached their inflection points in September 2001 and June 1996, respectively, implying that the growth rate of their PC development decreased successively.

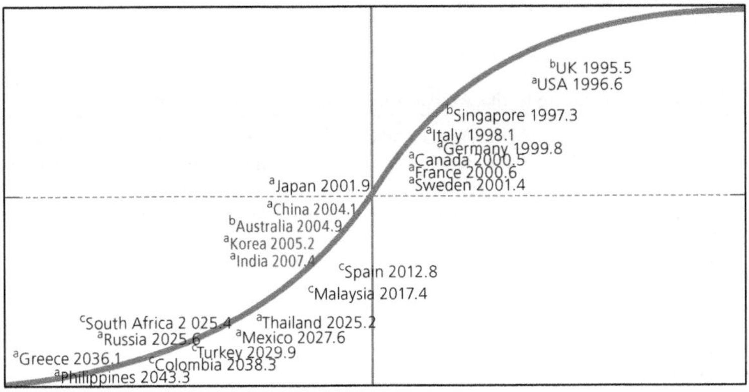

Fig. 9. PC Development Stage in 24 Countries (2002)

4.1.2 Sources of Increased Growth

The increase of China's PC growth (π) can be attributed to the multiplier effects of a phased development involving technology spillover effects, market learning, and efforts to catch up with international competitors.

Table 2 illustrates the rapid growth of China's PC development (referred to as 'activation' in the remainder of this section) which is attributable to the assimilation of technology spillovers, effects of market learning, and efforts to catch up with international competitors at a high level of statistical significance.

Based on this analysis, the contribution of respective factors to the activation of China's PC development from 1981 to 2002 is summarized in Table 3 and illustrated in Fig. 10. Both Table 3 and Fig. 10 indicate that, while mutual effects

Table 2. Factors Determining the Activation of China's PC Development

$$\ln\pi = -1.18 + 1.57 * \delta(t) + 0.97 \ln\lambda(t_{-1}) + 0.16 \ln D(t) + 0.45 D_{99-2001}$$
$$(-7.29)\ (81.7)\qquad (5.37)\qquad (19.84)\qquad (5.28)$$
adj. R^2 0.976 *DW* 1.46

where π: degree of activation of PC development; $\delta(t)$: assimilation absorption of technology spillover; $\lambda(t)$: effect of market learning; $D(t)$: effect of effort to catch up with international competitors; and $D_{99-2001}$: dummy variable (1999- 2001=1, other years=0).

of spillover, market learning, and efforts to catch up with international competitors accelerated the activation of China's PC development, the activation depended primarily on the effects of spillover and market learning during the imported PC components-dependent and market expansion periods. PC development in China continued to progress in its effort to catch up with international PC manufacturers in the competitive market period.

Table 3. Factors Contributing to the Activation of China's PC Development

	Change in rate of degree of activation ($\Delta\pi/\pi$)	Spillover (δ) and market learning (λ)	Catching up with international competitors (D)
Imported PC components-dependent and market expansion period (1981–1997)	7.7%	6.4%	1.3%
Competitive market period (1998–2002)	23.2%	6.5%	16.7%

The results of these analyses corroborate the hypothesis that "the rapid increase in the growth of PC development in China can be attributed to the multiplier effects of its phased development."

4.2 Coevolution of Rapidly Growing PC Development and Institutional Characteristics in Enhancing PC Innovation

4.2.1 Measurement of Institutional Characteristics

In order to identify the institutional features that have played a role in China's ICT development and utilization, 11 indices corresponding to the three dimensions of institutional systems as illustrated in Fig. 8 were selected, to facilitate a comparison with other countries (as analyzed in Section 3).

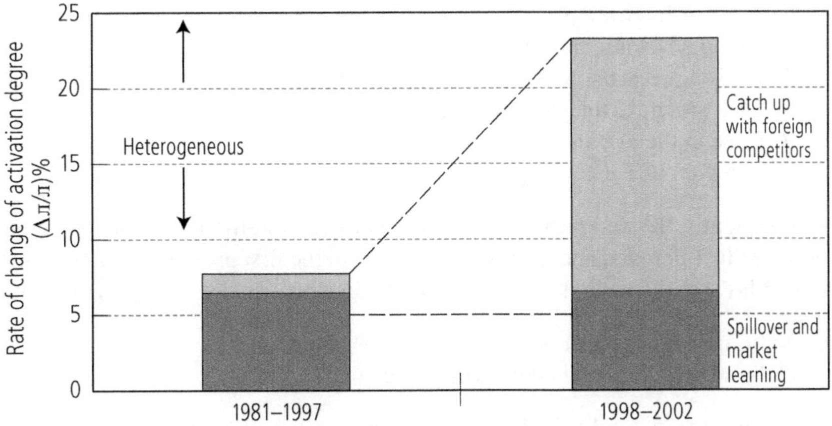

Fig. 10. Factors Contributing to the Activation of China's PC Development

A Principal Component Analysis (PCA) was carried out, using standardized data of 11 components of 24 countries for the year 2002. The descriptive statistics of principal components are summarized in Table 4.

Table 4. Descriptive Statistics of Principal Components

	Eigen value	Variance proportion (%)	Cumulative variance proportion (%)
PC1	6.55	64.9	64.9
PC2	1.78	12.7	77.6
PC3	1.14	10.0	87.6
PC4	0.38	3.4	91.0
PC5	0.32	2.9	93.8
PC6	0.31	2.5	96.3
PC7	0.24	1.6	97.9
PC8	0.13	0.9	98.8
PC9	0.07	0.5	99.3
PC10	0.04	0.4	99.7
PC11	0.03	0.3	100

Table 4 indicates that the first principal component has an extremely high level of Eigen value with a 64.9% variance proportion which increases to a 77.6% cumulative proportion by including the second principal component.

Weight and loading variables of these two selected principal components are summarized in Table 5.

1. A higher level of weight and loading of variables in the first principal component include:

a. Number of Internet users,
b. Enabling environment for ICT.
   Table 5 suggests the following important implications with regard to the institutional structure relevant for ICT development:
c. ICT use in the private sector, and
d. Nationwide ICT use.

2. Consequently, the first principal component can be interpreted as the discrepancy between ICT demand and usability. Contrary to the first principal component, the higher level of the weight of variables in the second principal component include:

   e. National culture (openness of the culture), and
   f. Cultural elasticity and adaptability to change.

3. Therefore, the second principal component can be interpreted as the openness and cultural elasticity toward the development and diffusion of ICT.
4. However, since the loading variables of the second component are smaller than those of the first principal component, the institutional structure determining ICT development seems to predominantly be influenced by the first principal component.
5. Thus, given that a nation's institutional structure is relevant for PC development as represented by the 11 previously identified indices, the structure depicted by the first principal component can be interpreted as the PC industry's institutional system, which reveals the discrepancy between PC demand and usability.

4.2.2 Institutional System Characteristics Relevant for PC Development

In Table 6 we present the differences in the PC industry's institutional systems (in terms of the gap between PC demand and usability) in 22 countries[6] by means of the principal component score for each respective country. Table 6 indicates that Indonesia has the lowest score (i.e., the discrepancy between PC demand and usability is highest) followed by China, India, Russia, Turkey, Philippines, and Thailand.

4.2.3 Effects of Activation of PC Development and the PC Industry's Institutional System Characteristics on the Functionality of PC Innovation

To identify the contribution of activation in PC development (π) to increased functionality of PC innovation—represented by PCs' elasticity to GDP (see Fig. 3) in PC developing countries—the correlation between them is assessed. Table 7 summarizes the results of the analysis, which demonstrates that a significant correlation

[6] Due to data constraints, South Africa and Colombia were excluded from the 24 countries in the preceding analysis.

Table 5. Weight and Loading of Variables in Selected Principal Components

	Standardized weight of variables in selected principal components		Loading of variables	
	PC1	PC2	PC1	PC2
Attainment of higher education	0.32	–0.11	0.83	–0.15
Nationwide ICT use	0.36	–0.15	0.92	–0.20
Number of Internet users per 1000 inhabitants	0.37	–0.05	0.96	–0.06
Use of ICT leveling public sector	0.31	0.16	0.80	0.21
Enabling environment for ICT	0.37	–0.003	0.96	–0.004
ICT use in private sector	0.37	0.11	0.95	0.15
Population	–0.18	0.09	–0.45	0.13
Ratio of people living in urban areas	0.31	–0.19	0.79	–0.25
Cultural elasticity and adaptability to change	0.11	0.64	0.28	0.86
National culture (openness of the culture)	0.07	0.68	0.19	0.91
Political instability and social violence	0.33	–0.09	0.86	–0.12

Table 6. Principal Component Score (PCS) of the First Principal Component: Differences in the PC Industry's Institutional System in 22 Countries (2002)

Country	PCS of the first principal component	Country	PCS of the first principal component
Indonesia	–4.46	Malaysia	–0.05
China	–3.52	France	0.62
India	–3.22	UK	1.51
Russia	–2.79	Germany	1.61
Turkey	–2.57	South Korea	1.74
Philippines	–2.30	Japan	1.88
Thailand	–2.07	Australia	2.80
Mexico	–1.77	Sweden	3.20
Greece	–1.11	Canada	3.35
Italy	–0.50	USA	3.72
Spain	–0.29	Singapore	4.22

exists between the activation in PC development and the functionality of PC innovation in 13 PC developing countries.[7]

Table 7. Correlation between the Activation in PC Development and Functionality of PC Innovation in 13 PC Developing Countries (2002)

$\ln \frac{\partial \ln V}{\partial \ln PC}$	$= -2.30 + 2.05\pi$
	$(-83.09)(14.52)$
adj. R^2	0.946

Fig. 11 illustrates this correlation.

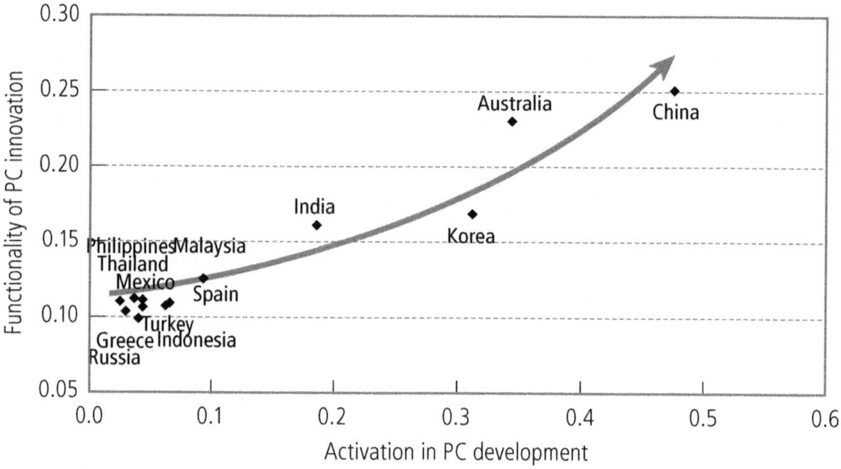

Fig. 11. Correlation between the Activation in PC Development and Functionality of PC Innovation in 13 PC Developing Countries (2002)

This analysis corroborates the second half of the first hypothesis that "China's activated PC development and diffusion induced an enhanced functionality of PC innovation."

Similarly, taking the principal component score of the first principal component (Table 5), which represents the differences in institutional systems relevant for PC development, the correlation between the characteristics of the PC industry's institutional system, and the functionality of PC innovation in 13 PC developing countries for 2002 is examined. The results of the analysis are summarized in Table 8.

[7] Although the size of the sample is limited to 13 (Tables 7, 8, and 9) and the leeway of these estimation results is fairly narrow, we tested the results in Tables 8 and 9 applying a heuristic approach without a constant dummy variable and obtained similar results. This verifies that the results of Tables 7, 8, and 9 are statistically significant despite the fairly narrow leeway.

Table 8. Correlation between the PC Industry's Institutional System and the Functionality of PC Innovation in 13 PC Developing Countries (2002)

$$\ln \frac{\partial \ln V}{\partial \ln PC} = -1.99 + 0.47DX + 0.13(1-D)X + 1.93D$$

$$(-31,02)(2.34) \quad (3.89) \qquad\qquad (2.54)$$

adj. R^2 0.657

Fig. 12 illustrates this correlation and reveals that the coefficients of China's, India's, and Indonesia's institutional systems are higher than those of the other countries. This demonstrates that the institutional system underlying PC development in these three countries accelerated the functionality of PC innovation more intensely than the other countries' institutional systems.

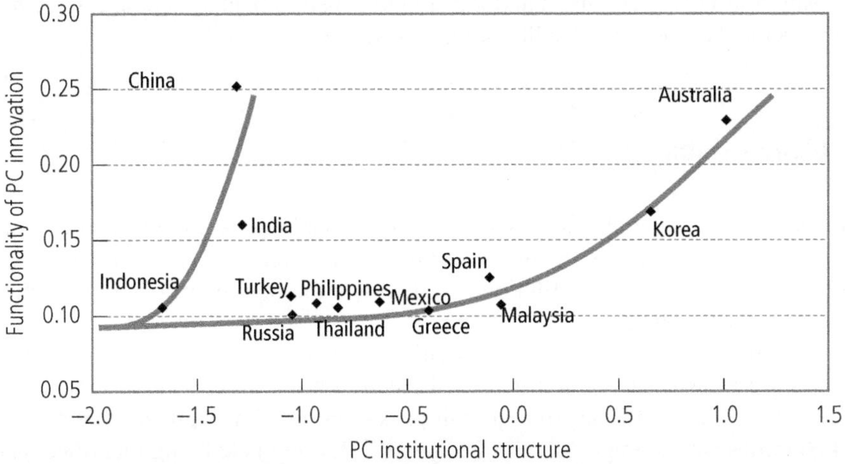

Fig. 12. Correlation between the PC Industry's Institutional Structure and Functionality of PC Innovation in 13 PC Developing Countries (2002)

4.2.4 Coevolution between Activation in PC Development and the PC Industry's Institutional System in Enhancing Functionality of PC Innovation

Prompted by the preceding analyses of the correlations between the significant impacts of both the activation in PC development and the PC industry's institutional system on the functionality of PC innovation, Table 9 demonstrates the correlation between activation in PC development, the PC industry's institutional system, and functionality of PC innovation in 13 PC developing countries.

Table 9. Correlation between Activation in PC Development, the PC Industry's System, and Functionality of PC Innovation in 13 PC Developing Countries (2002)

$$\ln\frac{\partial \ln V}{\partial \ln PC} = \underset{(-75.76)}{-2.29} + \underset{(10.60)}{1.68\pi} + \underset{(2.85)}{0.16DX} + \underset{(0.05)}{0.001(1-D)X} + \underset{(3.01)}{0.68D} + \underset{(3.70)}{0.24D_0}$$

adj. R^2 0.979

π: activation of PC development and diffusion; X: the PC industry's institutional system; D: coefficient dummy variable: China, India, Indonesia = 1, other countries = 0; D_0: constant dummy variable: Australia = 1, other countries = 0.

Table 9 denotes that both factors significantly contribute to the enhancement of functionality of PC innovation in China, India, and Indonesia, while the PC industry's institutional system does not contribute substantially in the other ten countries. This analysis corroborates the second hypothesis that "Coevolution between activation in PC development and a distinctive institutional system accelerated the enhancement of the functionality of PC innovation in China."

5 Conclusions

Given the distinctive characteristics of China's institutional system and the striking expansion of China's PC industry, this chapter aimed to determine the role of coevolution between the institutional system and the rapidly growing PC sector in fostering such growth.

First, in consideration of the possibility to model PC development trajectories as logistic growth curves and that the inflection point on this curve indicates the phase in which the PC sector's growth peaks, the PC development trajectories of 24 countries were compared, with China's PC development being identified as the fastest growing in the world.

Second, in line with our observation of the mutual causal relations between urbanization, higher income, higher education, and dependency on PCs underlying ICT's distinct institutional system in China, the ICT institutional characteristics of 22 countries were analyzed. To arrive at results, a principal component analysis was carried out, involving 11 institutional factors linked to the three dimensions of institutional system's impacts—including a country's national strategy and the socio-economic system, entrepreneurial organization and culture, and historical perspective.

On the basis of the empirical analysis of China's PC development and diffusion trajectory with a focus on its particular institutional system, the following hypothetical view on China's institutional structure that led to the rapid growth of PC development on a new level emerged:

1. The multiplier effects of a phased development (technology spillover, market learning, and high functionality development) have contributed to the rapid

increase of PC development and diffusion, inducing enhanced functionality with regard to PC innovation, and
2. The coevolution between rapidly growing PC development and China's distinct institutional system has accelerated the enhanced functionality of PC innovation.

These noteworthy findings provide new insights for further studies on institutional innovation.

Furthermore, a principal component analysis of the institutional structure of ICT was conducted with a focus on 22 countries, an approach that presents a new way to analyze complex systems. It would be particularly useful to conduct an in-depth analysis of the elements underlying these variables. This would help us understand the driving factors of effective ICT development.

Future research could explore additional institutional systems and characteristics of the ICT industry, which would provide significant insights and a better understanding of the interplay at work here.

References

Caniels M (2000) Knowledge spillovers and economic growth: Regional growth differentials across Europe. Edward Elgar, London

Douangngeune B, Hayami Y, Godo Y (2005) Education and natural resources in economic development: Thailand compared with Japan and Korea. Journal of Asian Economics 16(2):179–204

Flood R L (1987) Complexity: A definition by construction of a conceptual framework. Systems Research 4:177–185

Janes F R (1995) Interactive management: Framework, practice and complexity. In: Ellis K, Gregory A, Mears-Young B, Ragsdell G (eds) Critical issues in systems theory and practice, Plenum Press, New York

Kennedy C (1964) Induced bias in innovation and the theory of distribution. Economic Journal 74(295):541–547

Lee J W (2003) Economic growth and human development in the Republic of Korea: 1945–1992 `http://hdr.undp.org/docs/publications/ocational-papers/oc24aa.htm`

Lei S, Kita M, Watanabe C, You Y (2009) An empirical analysis of the institutional system's effects on the development of China's personal computer industry. In: Geenhuizen M v, Watanabe C, Jauhari V, Masurel E (eds) Technological innovation across nations: Coevolutionary developments, Springer, Heidelberg

Meyer P S, Ausbel J H (1999) Carrying capacity: A model with logistically varying limits. Technological Forecasting and Social Change 61(3):209–214

Nelson R R, Nelson K (2002) Technology, institutions, and innovation systems. Research Policy 31:265–272

North D C (1994) Economic performance through time. The American Economic Review 84(3):359–368

Oren S S, Smith S A (1981) Critical mass and tariff structure in electronic communications markets. Bell Journal of Economics 12(2):467–487

Phelps E S (1966) Models of technical progress and the golden rule of research. Review of Economic Studies 33:133–145

Teece D J (1986) Profiting from technological innovation: Implications for integration, collaboration, licensing and public policy. Research Policy 15:285–305

UNESCO (1998) World Culture Report 1998: Culture, creativity and markets, UNESCO Publishing, Paris

Watanabe C (2005) Institutional MOT: Co-evolutionary dynamism of innovation and institution. Proceedings of IAMOT 2005, Vienna

Watanabe C, Kondo R (2003) Institutional elasticity towards IT waves for Japan's survival. Technovation 23(4): 307–320

Watanabe C, Hur J Y, Matsumoto K (2005) Technological diversification and firm's techno-economic structure: An assessment of Canon's sustainable growth trajectory. Technological Forecasting and Social Change 72(1): 11–27

Yang J M, Kim T W, Han H O (2006) Understanding the economic development of Korea from a co-evolutionary perspective. Journal of Asian Economics 17(4):601–621

Weizsaecker C C (1966) Tentative notes on a two sector model with induced technical progress. Review of Economic Studies 33:245–251

Appendix: PC Development Trajectory in 24 Countries (1981–2002)

China

Parameter	Estimate	t-value	adj. R^2
K	8770	4.50	0.997
a	0.39	15.21	
b	–9.10	–34.57	Year
$-b/a$	23.07		2004.1

Japan

Parameter	Estimate	t-value	adj. R^2
K	8845	6.89	0.994
a	0.24	13.90	
b	–5.06	–36.82	Year
$-b/a$	20.87		2001.9

USA

Parameter	Estimate	t-value	adj. R^2
K	19384	4.17	0.990
a	0.19	7.11	
b	–2.89	–18.02	Year
$-b/a$	15.58		1996.6

France

Parameter	Estimate	t-value	adj. R^2
K	2795	1.95	0.996
a	0.18	8.19	
b	–3.53	–8.51	Year
$-b/a$	19.58		2000.6

UK

Parameter	Estimate	t-value	adj. R^2
K	2238	14.02	0.985
a	0.23	18.92	
b	–3.28	–51.26	Year
$-b/a$	14.50		1995.5

India

Parameter	Estimate	t-value	adj. R^2
K	3410	3.51	0.999
a	0.30	26.56	
b	–7.80	–52.57	Year
$-b/a$	26.40		2007.4

Sweden

Parameter	Estimate	t-value	adj. R^2
K	950	3.73	0.997
a	0.23	2.30	
b	–4.70	–12.07	Year
$-b/a$	20.43		2001.4

Spain

Parameter	Estimate	t-value	adj. R^2
K	4092		0.995
a	0.16	11.10	
b	–5.12	–42.08	Year
$-b/a$	31.78		2012.8

Greece

Parameter	Estimate	t-value	adj. R^2
K	55077	2.75	0.977
a	0.18	6.41	
b	–10.14	–0.32	Year
$-b/a$	56.33		2036.1

Colombia

Parameter	Estimate	t-value	adj. R^2
K	4373		0.810
a	0.10	2.65	
b	–5.60	–15.82	Year
$-b/a$	56.00		2038.3

Singapore

Parameter	Estimate	t-value	adj. R^2
K	214	23.60	0.992
a	0.45	12.58	
b	–7.26	–14.64	Year
$-b/a$	16.27		1997.3

Malaysia

Parameter	Estimate	t-value	adj. R^2
K	2431		0.946
a	0.16	3.12	
b	–5.68	–5.42	Year
$-b/a$	35.50		2017.4

Indonesia

Parameter	Estimate	t-value	adj. R^2
K	21172		0.933
a	0.25	9.89	
b	–9.37	–18.20	Year
$-b/a$	37.13		2018.1

Germany

Parameter	Estimate	t-value	adj. R^2
K	4649	7.42	0.989
a	0.19	11.66	
b	–3.60	–32.86	Year
$-b/a$	18.83		1999.8

Italy

Parameter	Estimate	t-value	adj. R^2
K	1332	6.09	0.994
a	0.25	12.69	
b	–4.35	–38.02	Year
$-b/a$	17.14		1998.1

Canada

Parameter	Estimate	t-value	adj. R^2
K	2370	2.15	0.998
a	0.15	7.87	
b	–3.07	–4.10	Year
$-b/a$	20.47		2000.5

Mexico

Parameter	Estimate	t-value	adj. R^2
K	10082		0.998
a	0.12	6.72	
b	–5.62	–7.50	Year
$-b/a$	46.60		2027.6

Russia

Parameter	Estimate	t-value	adj. R^2
K	14407		0.998
a	0.14	6.75	
b	–6.16	–14.18	Year
$-b/a$	44.60		2025.6

South Korea

Parameter	Estimate	t-value	adj. R^2
K	4005	2.96	0.992
a	0.23	11.20	
b	–5.48	–31.46	Year
$-b/a$	23.83		2005.2

Thailand

Parameter	Estimate	t-value	adj. R^2
K	6161		0.991
a	0.15	6.29	
b	–6.63	–25.57	Year
$-b/a$	44.23		2025.2

Turkey

Parameter	Estimate	t-value	adj. R^2
K	6963		0.880
a	0.13	6.02	
b	–6.13	–6.57	Year
$-b/a$	47.15		2029.9

South Africa

Parameter	Estimate	t-value	adj. R^2
K	4535		0.919
a	0.12	6.06	
b	–5.54	–14.33	Year
$-b/a$	46.17		2025.4

Australia

Parameter	Estimate	t-value	adj. R^2
K	2370	2.15	0.998
a	0.13	7.87	
b	–3.07	–4.10	Year
$-b/a$	23.90		2004.9

Philippines

Parameter	Estimate	t-value	adj. R^2
K	7994		0.998
a	0.10	6.63	
b	–6.26	–20.97	Year
$-b/a$	62.30		2043.3

[a]PC development trajectory is estimated by the following equation:

$$PC_T = \frac{1}{1+e^{-at-b}}$$

where PC_T: total number of PCs; K: carrying capacity; t: time trend; and a,b: coefficients. [b] $-b/a$: indicates the inflection point ($t = 0$ in 1981), Year indicates the year of inflection determined by $-b/a$ + 1981.

Institutional Context for IT Use in the Automotive Industry: A Case Study on the Market Leader in India's Passenger Vehicle Sector

Vinnie Jauhari

Abstract This study explores the use of IT in the automotive industry. It examines the institutional context of India's passenger vehicle sector by conducting a value chain analysis of the automotive industry. Additionally, the usage of IT at Maruti Udyog, India's market leader in the passenger vehicle sector, is investigated. The study evaluates to what extent IT has contributed to Maruti Udyog's market growth in India and in other, similar countries.

Key words: Automotive Industry, Information Technology, Institutional Influences, Value Chain Analysis, Maruti Udyog, Passenger Vehicle Sector, India

1 Introduction

This study explores the use of IT in India's automotive industry, specifically, in the passenger vehicle sector. A literature review illustrates the use of IT in the automotive industry and how companies can evolve from using IT for simple tasks to more complex ones—such as financial and consumer applications—which lead to costs savings and initiate strategies that usher in organizational change. The automotive industry's global context is examined and a value chain analysis conducted, which delineates measures that can lead to future sustainable development. Issues of importance for automobile manufacturers are highlighted and a general analysis of India's passenger vehicle sector is carried out. Subsequently, a case study evaluates the use of IT by India's largest passenger car manufacturer (Maruti Udyog Ltd) to demonstrate that even simple products, such as offering value-added services (e.g., after-sales support), setting up an extensive distribution network, and developing a real-time supply chain network, have led to the emergence of a cost-effective

Vinnie Jauhari
Region Lead, HP Labs Open Innovation Office, Tower D, 8th Floor, Mehrauli-Gurgaon Road, Gurgaon 122002, India, e-mail: vinnie.jauhari@hp.com

M. V. Van Geenhuizen et al., (eds). *Technological Innovation Across Nations*,

DOI: 10.1007/978-3-642-00158-1_3,

product line, making Maruti Udyog a market leader. The use of IT has been instrumental in reducing costs and making passenger cars available for the mass-market in India. In the five years since its inception, Maruti has been able to secure 75% of India's passenger vehicle market. This study concludes with the challenges the automotive industry faces and identifies measures that could facilitate this market sector's future growth in India, as well as in other, similar countries.

2 Global Scenario

The automotive industry is an important sector in the global economy, generating a considerable percentage of employment. For example, in the USA about 15% of the workforce is employed in the automotive industry (Capitaline Plus Database, 2006). The global car industry is growing at a rate of 2% per annum and is expected to continue growing at the same rate in the near future. Table 1 provides data on the production of passenger vehicles for various global regions.

Table 1. Passenger Vehicle Production by Region (x 1,000)

	1997	1998	1999	2000	2001	2002	2003	2004
North America	7,826	7,630	8,258	8,372	7,155	7,348	6,625	6,359
South America	2,044	1,607	1,355	1,650	1,722	1,672	1,642	1,992
European Union	13,377	14,464	14,844	14,779	14,939	14,741	14,602	14,664
Other Europe	2,178	2,231	2,448	2,628	2,435	2,496	2,634	3,158
Japan	8,491	8,056	8,100	8,359	8,118	8,618	8,478	8,720
South Korea	2,308	1,625	2,362	2,602	2,471	2,651	2,768	3,123
Other Asia-Oceania & Africa	1,737	1,650	2,394	2,825	2,987	3,832	5,219	6,212
Total	37,961	37,263	39,760	41,216	39,826	41,358	41,969	44,228

Source: www.autoindustry.co.uk/statistics, retrieved on April 18, 2006

Table 2 indicates the share of exports from different regions of the world. The European Union (EU) is the leading manufacturer and exporter of automotive products. This implies that the institutional context in EU member states is quite progressive and that technological advancements, coupled with a better understanding of the global market, have resulted in higher exports.

3 Value Chain Analysis of the Automotive Industry

Manufacturing sectors, such as the automotive or pharmaceutical industries, are driven not only by sales, but also by services offered together with the given product. That is, firms in these sectors move downstream toward the customer, aiming

Table 2. Leading Exporters of Automotive Products (in billion dollars and percentage)

	Value	Share in world exports				Annual percentage change			
	2002	1980	1990	2000	2002	1995–00	2000	2001	2002
European Union	303.30	52.8	53.8	46.8	48.8	3	–1	2	10
Extra exports	101.57	19.5	14.3	14.5	16.4	4	9	5	16
Japan	92.51	19.8	20.8	15.3	14.9	2	6	–9	15
United States	67.09	11.9	10.2	11.7	10.8	5	7	–6	6
Canada	56.33	6.9	8.9	10.5	9.1	7	0	–9	2
Mexico	30.91	0.3	1.5	5.3	5.0	17	18	0	1
Republic of Korea	17.30	0.1	0.7	2.6	2.8	11	17	2	12
Czech Republic	6.40			0.8	1.0	25	13	19	16
Hungary	5.98	0.6	0.2	0.8	1.0	49	1	12	12
Poland	5.19	0.6	0.1	0.7	0.8	32	80	6	23
Brazil	4.98	1.1	0.6	0.8	0.8	10	21	3	3
Turkey	3.16	0.0	0.0	0.3	0.5	19	5	54	35
Thailand	2.88	0.0	0.0	0.4	0.5	38	37	11	8
Slovakia	2.81			0.4	0.5	47	30	–5	23
China	2.68	0.0	0.1	0.3	0.4	19	10	–13	62
South Africa	2.40	0.1	0.1	0.3	0.4	19	10	–13	62
Above 15	603.92	94.4	97.1	97.0	97.3	–	–	–	–

Source: WTO (2003)

to tap into the profitable economic activity that takes place throughout the product life cycle. Table 3 depicts the various dimensions that can potentially contribute to an automobile company's revenue base. Services such as financing, insurance, fuel costs, and used vehicle purchase and sales are intertwined and run in parallel with the manufacturing process. Most large automobile manufacturers now offer such services, thus considerably increasing the significance of the service dimension in motor vehicle manufacturing. IT usage is not only confined to the manufacturing process itself, but is also found in various related areas.

Ford Motor Company, for example, offers onsite financing services through Ford Credit—the world's largest automotive finance company. Ford Credit assists over 4 million customers annually in financing their purchased vehicle. Ford's online scheduling service, the Virtual Service Advisor, is available around the clock for setting up vehicle maintenance and service appointments. Factory-trained technicians use the latest technology and state-of-the-art equipment and regularly enhance their skills through training offered at Ford facilities and the Ford Star satellite network (Ford Motor Company, 2006).

The automotive industry is a good example of a producer-driven chain (Humphrey, 1995). A producer-driven commodity chain refers to those industries

Table 3. Factors Contributing to Overall Costs throughout an Automobile's Longevity

Expenses	Percentage of total costs throughout the automobile's longevity
New car purchase	20%
Used car purchase	20%
Fuel	20%
Insurance	10%
Repairs	10%
Financing	10%
Other expenses	10%

Source: Adapted from Wise and Baumgartner (1999)

in which transnational corporations, or other large integrated industrial enterprises, play a central role in controlling the production system, including backward and forward linkages. This is the most important characteristic of capital- and technology-intensive industries producing manufactured goods like automobiles, computers, aircraft, and electrical machinery. According to Gereffi, "what distinguishes 'producer driven' production systems is the control exercised by the administrative headquarters of the TNCs." (Gereffi, 1994, p. 97).

In developed countries, about 60–70% of a vehicle's components are produced in-house. Automotive assemblers are in charge of the design process, distribution, consumer financing, and coordination of the used car market. Currently, the following shifts are being observed:

- The distribution system is being challenged by independent dealers;
- Western firms are increasingly outsourcing elements of their vehicle manufacturing processes;
- Leading suppliers are taking a more active role in the design process;
- Some design activities are being contracted out to independent design houses;
- Tier I automotive suppliers that supply original equipment manufacturers (OEMs) have assumed a key role and have entered the realm of product design; and
- Global models are increasingly being consolidated.

Two distinct paradigms for technological development can be discerned. One of the paradigms involves developed countries such as the UK, USA, Germany, and Japan with basic expertise in product design and an understanding of consumer dynamics which has been incorporated into the manufacturing process. The other one comprises developing countries, like Brazil and India, which have protectionist policies. The liberalization in India in the 1990s and in Brazil in 1995 resulted in a major revision in the countries' automotive policies, dismantling the system of control over investment and production. Global car manufacturers sought to position themselves in the developing countries' growing automobile markets, which enabled the dispersal of development costs, provided cheap production sites for the manufacture of selected vehicles and components, and opened new markets for higher end vehicles produced in industrially developed economies.

Several significant shifts in the pattern of suppliers are evident as well. Previously, relations between manufacturers and suppliers were necessarily close (Humphrey, 1999). Complex functions, such as design capability, largely remained in-house, while the subcontractors' and catalog suppliers' role was limited to delivering parts that were easy to produce (Helper, 1993). Catalog suppliers were at an advantage, since their production process was not customized (Laigle, 1995). New supplier networks evolved and supplier network models specific to each country were established. Black box manufacturing is also on the increase. Developing countries are being integrated into global production systems and global component suppliers have emerged, who are increasingly responsible for designing and delivering component systems to multiple worldwide locations. Relations between manufacturers and suppliers have also changed. Specifically,

a. Catalog suppliers and subcontractors have begun establishing relations with their customers and have moved toward black box manufacturing;
b. Designs are being developed to meet specific functional objectives;
c. Mergers and acquisitions in the field of design have taken place;
d. Shifts toward the supply of complete units rather than individual components are evident;
e. Tier I automotive suppliers now assemble complete units such as dashboards, rear axle assemblies, seats, etc.;
f. Manufacturing is moving from the automotive assembler to the first tier;
g. The involvement of automotive assemblers in the specification of their suppliers' production and quality systems has increased;
h. Suppliers have assumed responsibility for managing the remaining value chain; and
i. Universal designs are being pursued.

India's automotive components industry produces a wide range of products. With a turnover of approximately $4 billion, the Indian automotive component and parts industry is rather small by global standards. According to the Automotive Component Manufacturers Association of India (ACMA), the automotive component and parts industry registered a turnover of $3.8 billion in 2001. Over 77% of the total turnover originated from the formal sector. India's automotive components and parts industry consists of 400 businesses in the organized sector and over 5000 in the non-organized sector, who supply the organized sector. The non-organized sector has a market share of close to 23% and predominantly provides replacement and low technology parts, such as valves, gaskets, sheet metal, etc. Over half of these businesses have an annual turnover of less than $70,000.

One significant global trend that has recently emerged is that OEMs are increasingly turning to integrated systems suppliers in preference to individual component suppliers, a cost-cutting measure in light of increasing competitive pressure, as well as a means to enhance flexibility. In other words, the components supply chain is being rationalized—OEMs are opting for integrated systems suppliers because they have greater control over them in terms of costs and quality. OEMs expect their component suppliers to: (1) provide integrated systems rather than individual components, (2) participate in the global automotive chain, (3) raise quality levels, (4) participate in research and development, and (5) monitor and absorb a greater share of the warranty costs. These requirements have considerable implications for the use of integrated IT systems, involving both the supplier network, as well as the customer base. Alliances between various global automobile manufacturers also depend on networks that facilitate communication and the integration of their structures.

As a result of the extensive consolidation of automotive component suppliers in order to reap the benefits of economies of scale, the automotive component and parts industry is undergoing intense tierization. Consequently, the influence of Tier I suppliers, who supply OEMs directly, has increased tremendously. In a bid to reduce costs, tierization within the automotive industry has intensified at the global level. OEMs deal with Tier I suppliers directly, who deliver integrated systems rather than individual components requiring assembly. Tier I suppliers assemble the components in-house in accordance with the OEM's specification, and maintain direct contact with Tier II and Tier III suppliers, who deliver subsystems or special components.

Tierization is popular in the American, Japanese, and European markets on account of their high sales volumes. These markets provide a reliable sales base for vendors and are well-integrated, making minimal taxation rates, as well as cost advantages for the supplier, possible. In contrast, sales in India are relatively low. Moreover, India's tax regimes are not aligned and cost advantages for the supplier are less likely. This may affect the domestic economy's pace of reform and, consequently, slow down tierization in India.

OEMs in India are also reducing the number of vendors for each particular automotive component. The country's largest commercial vehicle manufacturer, TELCO, has reduced the number of vendors from 1,200 to 700 and plans to further decrease the number to 450. Bajaj Auto slashed the number of vendors from 1,400 in March 1998 to 400 by December 2001. Vendor rationalization not only reduces the costs associated with multiple vendor transactions, but facilitates the process of quality control and enhances the efficiency of supply chain management. Furthermore, the OEMs' increase in order size per vendor broadens the vendors' scale of operations, which in turn enhances their efficiency. Thus, incentives for vendors to implement technological improvements and develop systems and processes that meet the quality standards set by OEMs are being provided.

4 The Use of IT in an Organizational Framework

Based on research findings by Gibson and Nolan (1974), as well as subsequent research by Nolan (1979), six stages in the development of electronic data processing (EDP) facilities within an organization can be identified, leading from the introduction of computers in the organization to efficient management of data processing. This path of development is illustrated in Fig. 1, which is based on the finding that companies' EDP budget frequently forms a double S-shaped curve when planned over a period of time, beginning with initial investment in EDP to advanced operations. The double S-curve underlying this pattern of EDP development consists of six stages—initiation, contagion, control, integration, data administration, and maturity.

Stage 1 Initiation: computers are introduced as a cost-saving measure and are met with ambiguity and with a resistance to change.

Stage 2 Contagion: additional computing facilities are acquired by the organization for more comprehensive and advanced applications, leading to a period of rapid expansion. Expenditures for hardware, software, and personnel rise sharply. It is a period characterized by 'contagious' unplanned growth and by increasing responsibilities of the EDP manager. Curbing the often chaotic effects of rapid development depends alone on the quality and judgment of the staff directly involved.

Stage 3 Control: as the IT budget increases, top management begins to replace analysts with more valuable technical staff. Investments in IT are compensated by cutting general and administrative expenses, i.e., replacing administrative staff with IT programs for tasks such as accounting. The focus shifts toward inventory control systems. Applications that have significant potential to increase revenues and profits and facilitate managerial decision-making have not yet been installed (e.g., financial planning models and on-line customer service systems). The reason why such applications have not been implemented to date is the trade-off dilemma between the centralization and decentralization of various functions of the EDP/MIS (Management Information Systems).

Stage 4 Integration: application development focuses on the establishment, utilization, and internal sharing of organizational databases, i.e., the integration of the organization. The focal shift from IT/computer use for improving the efficiency of operational systems to data processing and database management opens up a new range of potential IT applications that to a large degree are transformational in nature.

Stage 5 Data Administration: demand for better control and efficiency increase. An organizational database is set up and maintained, and design applications for the effective utilization of the database are developed.

Stage 6 Maturity: the application portfolio's structure reflects the organization and information flows within the company. The computer and data resources reach a level of maturity within the organization.

The global automotive industry has reached the maturity stage. Automakers in India are still at various stages of development in accordance with Nolan's framework. Maruti Udyog, however, is an exception and can, in contrast to its national competitors, be classified as having reached the maturity stage.

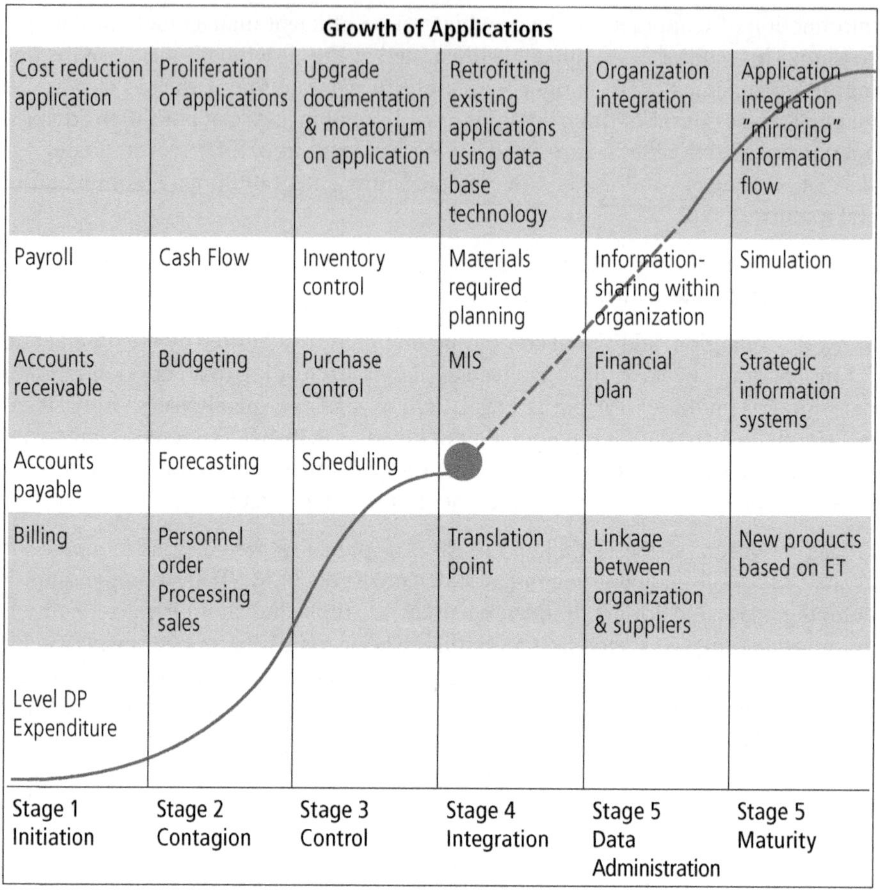

Fig. 1. The Use of EDP Facilities within an Organization
Source: Kanungo (1999)

An alternative method of ascertaining IT use in an organizational context is the life cycle approach. An automobile life cycle consists of the following generic stages (Carli et al., 2006): materials production, manufacturing and assembly, use and service, and end of life management. Key stakeholders throughout the life cycle include material suppliers, parts fabricators, professionals, OEMs, customer service and repair professionals, dismantlers, shredders, non-ferrous processors, waste managers, and regulators. IT offers distinct advantages throughout this life cycle, for example, the streamlining of automobile manufacturers' operations. Some companies have set up corporate intranets and share illustrations of competitors' products, thus making a visit to the competitors' vehicle teardown areas superfluous. Other firms use the Internet to communicate directly with their customers, who have the possibility of placing customized orders and eliminating the need of an intermediary. More recent IT activities include investments in private extranets to link multiple trading partners and in trading exchanges that connect buyers and suppliers to more effectively design and market products, manage supplies, and provide customer service (Mcginn, 2002).

King and Lyytinen (2004) assessed the role of information technology and its impact on enterprise transformation in the automobile industry. Their study traces the historical role of IT in the automotive industry, predicts some future transformations, and explicates an analytical strategy to examine transformations resulting from IT use. Within their framework of assessment, King and Lyytinen categorize the automotive industry as including "all of the activities and infrastructure required to make automobile transportation possible. Beyond OEM and sales, this includes all of the tiers of supply in original production, and the servicing of automobile transport thereafter (fuel, insurance, streets/highways, regulation, resale, disposal). Information technology includes all modes of information collection, processing, storage, and dissemination. In addition to modern digital technologies, this includes manual methods that have been part of the industry since the earliest days" (p. 286). The impacts of IT on the automotive industry are summarized in Table 4. Accordingly, IT supports organizations by linking them with the supply chain, manufacturing, logistics, and the distribution networks, in addition to offering other value-added services.

The literature provides some illustrative examples of IT use. For example, Volvo developed the Lambda sond system in 1970, which consisted of an exhaust gas composition sensor, a feedback loop to the fuel injection system, and the catalytic converter. The system monitors oxygen levels in the exhaust gases and forwards the results to an electronic unit which resets the fuel injectors as needed. This was the first application of modern information technology to monitor and control an automobile engine's emissions. The sond system was incorporated into Volvo's 240 models sold in California in 1976. The introduction of emission control systems had cost implications for consumers, since the equipment eventually wore out and the corroded parts had to be replaced. Subsequently, regulators ordered OEMs to introduce five year or 50,000 mile warranties on all vehicle components connected to the emissions control system.

Toyota, for example, has invested in innovations to reduce gas emissions and improve fuel efficiency. The company has made efforts to reduce hydrocarbon, carbon monoxide, and nitrogen gas emissions through Diesel Clean Advanced Technologies—which incorporates technologies such as a computer controlled common rail fuel injection system—or the new Diesel Particulate Nox Reduction System (DPNR), which removes particulate matter (PM) and Nox from exhaust emissions. Toyota has extended the scope of its value chain to construct an integrated customer lifestyle support system. The main elements of the value chain Toyota focuses on include e-commerce, financing, and Intelligent Transport Systems. The company's e-commerce venture in Japan is steadily growing with Web site membership (gazoo.com) having reached three million. On average, 360,000 online requests for price quotes and catalogs are received by Toyota dealers annually, with 10% of these requests resulting in purchases of Toyota vehicles. The Toyota Financial Service Corporation network has expanded to cover 22 countries and is planning to

Table 4. Regimes of IT Impact on the Automotive Industry

Regimes of IT impact	Product, process, or regulation regimes	Examples of IT application and impacts
Use and service monitoring	Property regulation, risk mitigation, and complimentary asset provision	Car and driver registration services, insurance
Enhanced product platforms	Atmospheric emissions control	Embedded digital fuel and emissions control system
	Passenger safety	Air bag and collision detection systems
	Entertainment, conviviality, and control	Sound systems, hands-free phones, remote control
Production and distribution systems	Expediting and coordinating production and distribution	MRP, ERP systems, supply-chain management (EDI, e-collaboration, e-markets)
	Manufacturer–customer relationship establishment and maintenance	Car service history, customer profiling/service, intelligent service management

Source: King and Lyytinen (2004)

establish additional networks. Toyota offers its own credit card, the TS Cubic Card, as well as one-stop financial services. Toyota's 'G Book system' is a comprehensive automobile information system—users connect to their respective Gazoo center and obtain information on various services via onboard terminals installed in their vehicle. Lee and Whang (2001) have analyzed the impacts of IT on the dynamics in the supply chain, and concluded that IT impacts on four key dimensions: information integration, planning synchronization, work flow coordination, and new business models (Table 5).

IT has been used extensively in the production and distribution of new vehicles and the re-marketing of used vehicles. It is also useful to take into account the long history of organizational innovation in the automotive industry. For example, Japanese OEMs developed techniques for quality assurance and lean manufacturing, one of the most important aspects being the *kanban* system, an innovative information management process for coordinating the production system. The *kanban* system fostered a Just-in-Time inventory control strategy which was eventually adopted by OEMs worldwide. The use of IT for digital computer and process control devices is well known. Robotics has been applied to repetitive production processes, and IT systems have been used in supply chain management processes, have linked suppliers to OEMs, and facilitated documentation and order entry. These factors have contributed to the matching of supply with demand.

The Covisint initiative by the 'Big Three' U.S. OEM suppliers (GM, Ford, and DaimlerChrysler) established a marketplace for auctions to match supply with demand. The three leading automakers developed Web sites which provide prospective buyers with information on product characteristics and costs that is useful for

Table 5. Impact of IT on Various Aspects of Work in the Automotive Industry

Dimension	Elements	Benefits
Information integration	Information-sharing and transparency; Direct and real time accessibility	Reduced bullwhip effect; Early problem detection; Faster response; Trust-building
Synchronized planning	Collaborative planning, forecasting and replenishment; Joint design	Reduced bullwhip effect; Lower cost; Optimized capacity utilization; Improved services
Workflow coordination	Coordinated production planning and operations, procurement, order processing, engineering change, and design; Integrated, automated business processes processes	Efficiency and accuracy gains; Fast response; Improved service; Earlier time to market; Expanded network
New business models	Virtual resources; Logistics restructuring; Mass customization; New services; Click and mortar models	Better asset utilization; Higher efficiency; Penetration of new markets; Creation of new products

Source: Lee and Whang (2001)

selecting a vehicle and negotiating a reasonable price. Lee and Amaral (2002) carried out a study on DaimlerChrysler's Mopar Parts Group which is a $4 billion U.S.-based OEM automotive parts and accessories distributor for dealerships in the USA and Canada. Mopar has a highly complex supply chain involving 3,000 suppliers, 30 distribution centers, and 225,000 orders per day from 4,400 North American dealers. Mopar's SCPM (Supply Chain Performance Management) system identifies performance exceptions by monitoring demand forecast, inventory, and supply performance metrics relative to predefined objectives. In the first year alone, DaimlerChrysler shortened its decision cycle time from several months to a few days, reduced excess transportation system costs, increased its order fill rate by 1 percentage point, and cut inventories by $15 million.

In addition, Lee and Amaral (2002) assessed potential *unintended* consequences related with disruptive supply chains. For example, late customer shipment could result in production disruptions and delays in critical orders. High material costs could lead to increased scrap and return rates and, consequently, to customer dissatisfaction, while poor incoming material quality could result in high inspection and obsolescence costs. Unmanageable semi-knocked units could entail lower product distinctiveness. The use of IT can significantly reduce such costs, as has been demonstrated at the global level.

To summarize, IT has a considerable impact on all stages of manufacturing and offers value-added services to customers during and after the purchase of their vehicle. IT can greatly enhance customers' purchasing experience. In a global context, where the challenges of recycling, fuel efficiency, alternate sources of energy, and safety mechanisms have become essential for sustainable competitive advantage, the use of IT can facilitate the shift toward a sustainable development paradigm. For the consumer, the greater variety of products in addition to the automaker's standard products and the convenience of after-sales services will increasingly become a decisive factor in the decision to purchase a car. The introduction of newer technologies will enhance the vehicle's longevity and, hence, its usage, resulting in higher expenditures on maintenance. Buy-back facilities will also have an impact on vehicle sales. IT could become an important interface for automakers in the establishment of customer networks, as well as networks between the various manufacturers. The trend toward outsourcing and tierization in the global supply chain indicates a shift toward increased IT application in the automotive industry to set up integrated supply systems and use common IT platforms for global buying and selling.

5 The Context of India's Automotive Industry

India's automotive industry had a modest start between 1910 and the 1920s with the establishment of assembly plants in Mumbai, Calcutta, and Chennai. In 1946, Premier Automobiles was licensed to manufacture the first passenger vehicles in India, Dodge, Desoto, and Plymouth cars, at its Kurla plant. Shortly thereafter, Hindustan Motors (HM), initially an automotive parts manufacturer, began assembling cars in 1949. In 1952, tariff regulations were imposed by the Indian government in its effort to devise regulations for the development of an indigenous automobile industry. Assembly plants that did not intend to set up their manufacturing facilities in India were required to shut down their operations. This marked the beginning of control measures in India's automotive industry. There was little competition and the restrictions on international collaboration resulted in limited technology import. The three decades following the establishment of the passenger vehicle industry were characterized by restrictive government policies and the existence of only two car brands on the market–Ambassador and Padmini. The 1980s ushered in a period of change in India's automotive industry.

In the early eighties, a series of liberal policy changes was announced, initiating a turning point in India's automotive industry. The Government of India entered into the automotive sector with a 74% stake in Maruti Udyog Limited (MUL), a joint venture with Suzuki Motors Ltd of Japan. Maruti Udyog launched the Maruti 800 as a 'people's car', making passenger cars affordable for India's middle class. Maruti Udyog was the market leader until the early 1990s, securing nearly 90% of the market share in the passenger vehicle sector. In 1991, the government announced a series of reforms which led to intensified competition in the sector. Automakers including Hyundai, Ford, Honda, Daewoo, Mitsubishi, Fiat, and Skoda entered the Indian market in the nineties and early 2000. Competition intensified with the increase in models available to the mass-market, and Hyundai's Santro and the Tata

Sierra, the first indigenous Indian passenger car manufactured by the Tata Group, began to make a dent in Maruti's market share. The market shares of the different passenger vehicle segments in India are illustrated in Table 6.

Table 6. Market Share (%) of Passenger Vehicles by Segment

	1998–1999	2000–2001
Economy–Segment A Maruti 800, Omni, Padmini Ambassador	62.2	40.1
Family–Segment B Hyundai Santro, Maruti Zen Wagon R, Maruti Alto, Tata Indica, Fiat Uno	26.1	46.2
Executive–Segment C Ford Ikon, Hyundai Accent, Maruti Esteem, Honda City, Mitsubishi Lancer, Maruti Baleno, Siena	9.9	13.6
Luxury–Segment D Mercedes, Toyota Camry	0.2	0.1

Source : Centre for Industry and Economic Research, Industrial Techno Economic Services P. Ltd

To sum up, the restrictive market environment before 1991 prevented strong competition and, hence, no indigenous Indian brands were available in the passenger car sector. Only the Tata Group manufactured a fully indigenous car, the Tata Indica, which was introduced in the late 1990s. Prior to 1991, there was little investment in the development of technology. Exports of passenger vehicles in the eighties and early nineties were insignificant, since Indian automakers faced capacity constraints that even hampered supply to the domestic market. However, the establishment of Maruti Udyog in the 1980s as a joint venture with Japanese collaboration stimulated investments in the automotive sector. A manufacturing plant was set up in collaboration with Suzuki Motor Company, exposing Maruti Udyog to the Japanese system of manufacturing, including the consolidation of a strong vendor base, the use of IT in manufacturing, Just-in-Time production, and the initiation of product- and process-based innovations. Maruti Udyog is built on a zero-inventory model and procures 80% of its auto parts from suppliers located in close proximity of the assembly plant. The automaker has made considerable investments in supply chain management. In fact, Maruti's investments in IT have been one of the main reasons for its low production costs and the attainment of mass-market status in India, thus increasing the company's sales volume. Since 2000, foreign direct investment (FDI) in India's automotive industry has led to the country's emergence as an outsourcing destination, due to lucrative cost advantages and its supplier quality systems.

Other institutions also played a role in fostering the automotive industry's progress. Up to 100% FDI is now permissible which has intensified competition considerably. The supplier base continued to develop and consumers were now offered a much wider range of vehicles to choose from. This, in turn, increased the pressure on automakers to produce state-of-the-art cars and forced them to consider other components underlying the value chain in order to increase revenue. This is in line with global trends, as observed at the Ford Motor Company, which, for example, now offers vehicle financing services to customers.

The opening of India's economy in 1991 led to increased entrepreneurship and more job opportunities with higher wages. In addition, liberalization measures in the banking and finance sector led to lower lending rates. This had a positive impact on the demand for cars and the automotive industry in general. A higher demand in all vehicle segments, as well as a shift toward higher end models, became evident. Increased disposable income, changes in lifestyle, the lack of public transport facilities, and improved availability of financing options are the key demand drivers in India's passenger vehicle sector.

Motor vehicle manufacturing basically involves the assembly of components procured from ancillaries or automotive component manufacturers. Nearly 80% of automotive component manufacturing is outsourced, which reduces the capital cost needed to set up a vehicle manufacturing plant. Therefore, automotive ancillaries play a key role in maintaining the quality and price of the product. Only after Maruti Udyog was established was supplier development considered part of the country's vehicle manufacturing operation. India was first introduced to the concept of outsourcing through the setting up of automotive component manufacturing facilities by Indian entrepreneurs, in collaboration with Japanese counterparts, to establish a supply base for Maruti.

With new market entrants planning to set up manufacturing facilities in India with a small capacity base, the role of automotive component suppliers will substantially increase in the next few years. Presently, some of the luxury automakers import CKD (Completely Knocked Down)/ SKD (Semi-Knocked Down) kits and carry out assembly operations in-house. However, the Indian government's strict policy guidelines contend that all manufacturers have to opt for a 70% indigenization within a period of five years following the launch of manufacturing operations in the country. This regulation will further boost the operations of automotive component manufacturers in India.

According to the Automobile Component Manufacturers in India (ACMA, 2006), India has the fourth-largest passenger vehicle market in Asia, the second-largest two-wheeler market in the world, as well as the fourth-largest tractor and fifth-largest commercial vehicle market worldwide. The vehicle market segments in India, with a total production volume of 8.46 million vehicles include: 14.5% passenger cars, 77% two wheelers, 4.5% three wheelers, and 4% commercial vehicles (ACMA, 2006). The growth pattern of Multi Utility Vehicles (MUVs) and of the passenger vehicle sector are provided in Table 7. In 2004–05, the passenger vehicle sector's growth rate was 22% higher than in the previous year. The projected size of the automotive component market by the year 2015 is estimated at $33–40 billion. After exporting 38,000 passenger vehicles to Europe in 2003, Suzuki planned to use Maruti as a production base for exports. With its high machine tool capability, improving tooling capability, exceptional automotive component industry, and the fact that most raw materials are available locally, India fulfills the requisites of an excellent manufacturing base. Together with its high quality, reasonable pricing, and special capability to supply smaller volumes, India is also an exceptional automotive component supplier (Capitaline Plus Database, 2006). Foreign automotive majors with a manufacturing base in India are rapidly expanding their indigenization programs. India has the potential to become a global sourcing base for auto components in the short term, for CKD/SKD in the medium term, and completely built units (CBUs) in the long-term. A number of foreign cars are manufactured in India, with 70% of these vehicles' components being produced locally. The most prestigious customers of Indian auto component exports include GM, Ford, Chrysler, BMW, Audi, Mercedes-Benz, Volvo, Isuzu, Nissan, Piaggio, and New Holland. Europe accounts for around 35% of total auto component exports from India, America follows with 25%, and exports to Asia and Africa total 17% and 13%, respectively. The majority of auto components exported are engines and engine parts (35% of all auto component exports), transmissions and steering parts (19%), suspensions and brakes (10%), and equipment (7%). Of a total of 402 companies, 240 are exclusively export-oriented, while 88 are OEM/Tier 1 exporters.

Table 7. Production of Passenger Vehicles, Components and Exports

Year	MUVs (x 1,000)	Passenger vehicles (x 1,000)	Auto component industry output (in $ million)	Auto component industry exports (in $ million)	Investments (in $ million)	Export share to output*
1994-95	49	264				
1995-96	67	348	2588	267	1315	
1996-97	134	407	3278	291	1705	
1997-98	134	401	3008	330	1813	
1998-99	113	390	3249	350	1850	
1999-00	123	574	3804	456	2000	
2000-01	127	513	4100	625	2300	
2001-02	105	564	4470	578	2300	13%
2002-03	114	608	5430	760	2645	14%
2003-04	146	842	6730	1020	3100	15%
2004-05	182	1028	8700	1400	3950	16%

*Components used in exported vehicles: $1,200 million
Source: Auto Component Manufacturers Association India, 2006

The government introduced the Auto Policy in 2002 with the objective of establishing a globally competitive automotive industry and double the industry's contribution to the economy by 2010. The government also pledged to establish an adequate regulatory framework meeting environmental norms. India now has the largest number of compressed natural gas (CNG) operated bus networks in the world. The conditions for the import of completely built units will be issued by the Director General of Foreign Trade (DGFT), taking environmental and safety regulations into account. The government aims to create an adequate fiscal structure to facilitate the production and sale of hybrid vehicles, but has yet to introduce a model to levy higher road taxes for older vehicles to comply with international standards.

6 IT in the Broader Perspective of Technology Development

Government support of research and development (R&D) activities in countries like the USA, South Korea, UK, and Germany has been quite impressive. The developed world accounts for around 95% of total global R&D spending (Forbes and Wield, 2000). Table 8 illustrates R&D investment as a percentage of select countries' gross domestic product (GDP).

Table 8. R&D as a Percentage of GDP (2005)

Country	2005
USA	3.5
Japan	3.2
China	1.4
Germany	2.5
Brazil	1
Hungary	0.9
India	1

Source: Adapted from Battelle (2005) Global R&D Report

Over 80% of industrial R&D is devoted to improving products that already exist (Rosenberg, 1996). Hobday (1995) argues that several firms in Taiwan, Singapore, South Korea, and Hong Kong transitioned from being original equipment manufacturers (OEM) to original design manufacturers (ODM) and are now original brand manufacturers (OBM). Forbes and Wield (2000) have introduced a possible pattern of growth for technology firms:

- Learning to produce;
- Learning to produce efficiently;
- Learning to improve production;
- Learning to improve products;
- Learning to develop new products.

The transition from being an OEM to becoming an ODM is challenging. Technology firms should focus on new design, which usually involves the incorporation of new components, materials, or manufacturing methods into an existing product or product process. Designs based on existing technology help the product move up the value chain. For example, Hyundai intentionally acquired style design capabilities in the 1970s as part of its automotive development plan. A team of five design engineers was selected to study automotive design and styling, and sent to Italy to participate in design processes, kept a close record of what they had learned during the day, and conducted group reviews every evening. These five engineers eventually made up the core of Hyundai's design department, one of whom headed R&D at Hyundai (Kim, 1997).

Kim (1997) has also tracked the efforts of Hyundai Motors of Korea, which mobilized substantial R&D resources to develop its engines, a typical case of 'catching up' made possible by huge investments. Hyundai was, by and large, left to its own devices in the development of its engines. Mitsubishi provided some assistance, but automakers were generally not very keen on sharing their technology, and Hyundai sought support from specialized R&D firms, like Ricardo in England. Hyundai's technological development can also be described as 'stage-skipping' to 'catch up'. When Hyundai began developing engines, carburetor-based engines were the standard engine types. As engine technology was developing toward a new electronic injection-based engine (Kim, 1994), Hyundai was able to reduce the gap in its engine technology within a very short period of time.

By contrast, technology in India has primarily been outsourced by means of global partnerships. Suzuki was instrumental in introducing the contemporary car in India in the 1980s, and soon became a market leader by offering additional value-added services. Maruti Udyog, on the other hand, was protected from competition for nearly seven years before the Indian government initiated liberalization measures. In its early years, Maruti focused on producing more cost-effective vehicles, rather than concentrating on developing indigenous models. No government or private initiative existed that promoted the development of a specifically Indian car brand. It was only in the nineties that Tata developed an indigenous car brand, the Indica, which is now exported as well. As a consequence of technology spillover from Maruti Udyog, efforts were undertaken to further expand the auto component industry. The supply chain met Maruti's real-time delivery schedules for most of the required automotive components. This also had an effect on other automakers that set up their operations in India after 1991.

7 Information Technology at Maruti Udyog

The advantages of information technology in the automotive industry have huge implications. From a broader manufacturing perspective, as described in the previous section, IT can be deployed in different functional areas (Fig. 2). This requires an open systems view of the organization in terms of information exchange, as illustrated below.

IT can act as an interface between all departments and at various levels. The broad environmental context in which IT is adopted may lead to disparities in IT use within the organization. Differences in national incomes, cultural values, infrastructure, the degree of market competitiveness, political position vis-à-vis technology, and the country's level of technological sophistication are just some of the factors that greatly influence the adoption of IT in all areas. At the organizational level, the performance of the company may depend on a number of factors rather than simply on the use of information technology.

Maruti Udyog is India's largest automaker, with a market share of over 80% in the national automotive industry. It was set up as a joint venture by the government

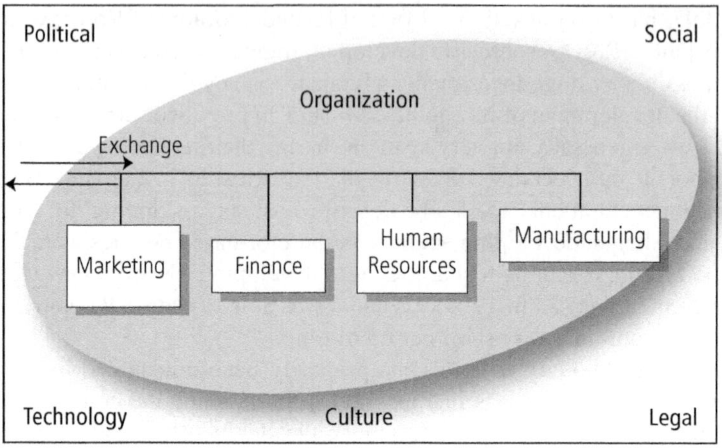

Fig. 2. Systems View of the Organization

of India and Suzuki Motor Corporation, a Japanese partner. Although the company is India's market leader, it is facing severe competition from new entrants in the automotive industry. Maruti Limited was established in June 1971 and went into liquidation in 1977. The company was incorporated as Maruti Udyog Ltd in 1981, and its assets acquired by the government under the Maruti Udyog Acquisition and Transfer of Undertakings Act of 1980. The purpose of Maruti Udyog's establishment was to develop an affordable passenger car for India's middle class, which featured contemporary design and quality. This fostered the development of India's automotive component industry, and local entrepreneurs were granted investments and long-term contracts. With Suzuki's assistance, new technology and management practices were introduced at Maruti. The first car, the Maruti 800, was released in 1983. By 1985, when an all-terrain vehicle was launched, Maruti had already sold 50,000 vehicles. Today, the company exports to over 50 countries, including, e.g., the Netherlands, Austria, and Belgium. Maruti is represented in 898 cities and ranked first in the Customer Satisfaction Index of the JD Power Survey. The company is known for its value-for-money pricing strategy which has been made possible by the high level of indigenization of its vehicles. Most models are indigenized to over 90% (though some exported models are indigenized up to 76%). Maruti's supplier network consists of about 375 vendors, and the automaker has several joint ventures with some of these to source its raw material requirements. With regard to sales and marketing, Maruti Udyog has a network of 112 dealers and sales outlets in 86 locations across the country. It also has a network involving 1,010 workshops in 412 cities and towns throughout the country.

Following the international trend in the automotive industry, Maruti offers a diverse range of services for consumers to increase their ownership experience. Divisions, such as Maruti Finance, Maruti Insurance, Maruti True Value and Maruti

N2N, Maruti Genuine Parts, and Maruti Genuine Accessories, have been introduced to satisfy customer demand for a broad variety of services.

7.1 Information Management at Maruti Udyog

Maruti Udyog's manufacturing operations are characterized by a very high degree of IT integration. It spans across supply chain management, HR, marketing, finance, and dealer networks. The structure of the company's IT division is indicated in Fig. 3.

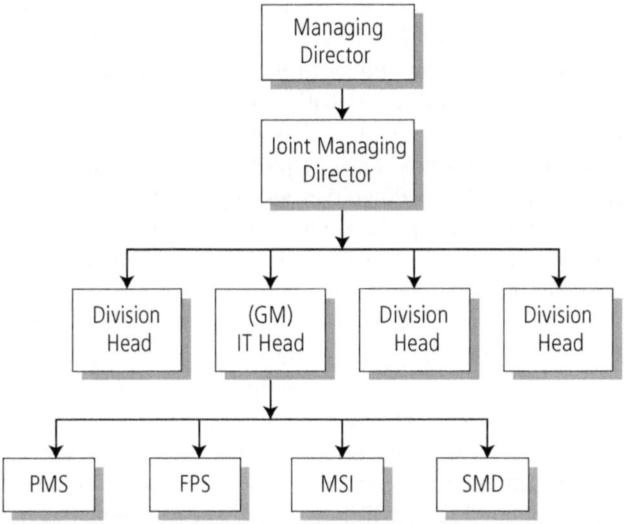

Fig. 3. IT Structure at Maruti Udyog
PMS: Plant Material System; FPS: Finance and Personnel Management; MSI: Marketing and Sales Information System; SMD: Systems Management and Development

The functions of the Plant Material System (PMS) are extensive and include vendor and supplier management, receipts and inventory management, material consumption, stock accounting, sales/excise tax, spare part sales, raw material procurement, and material requirements planning (MRP). The Finance and Personnel Management system (FPS) is in charge of all matters regarding financial management and HR-related aspects, such as manpower, compensation, etc. The Marketing and Sales Information system (MSI) is involved in marketing, sales, and customer service. Finally, Systems Management and Development (SMD) monitors Maruti's IT infrastructure and ensures functional support at all times. The entire manufacturing cycle consists of two parallel processes, as illustrated in Fig. 4.

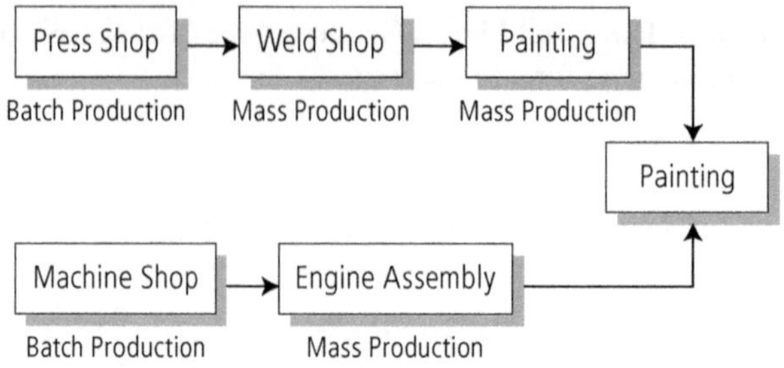

Fig. 4. Production Process at Maruti Udyog

All of Maruti Udyog's materials handling processes are automated. Annual manufacturing plans are formulated by the top management in consultation with the Managing Director. Production is broken down into daily, line, and model perspectives which, when multiplied with aggregated material, provides the total gross material requirement for the day. This figure is then compared with the running material block to determine daily requirements. The suppliers are connected to this system online and can download and upload the information around the clock. Monthly manufacturing plans are broken down into the following categories: hourly-based supply system, delivery instruction system (daily-based supply system), and months-based supply system.

Maruti's inventory is stocked for less than one day. It has a real-time supply chain and, hence, inventory pile-up is avoided, resulting in tremendous savings. The waiting period for supplying inventory stock for auto components that are imported is approximately five months. With an indigenization of 90-95% (despite the fact that some car models are 76% indigenized), Maruti's cost savings are substantial. 70% of Maruti's suppliers are located within a periphery of about 125 miles from the manufacturing plant. The remaining 30% are scattered across the country. For these suppliers, requests are sent ten days in advance. Maruti has invested heavily in training and developing the skills of its employees, and in upgrading its technology for more efficient supplier management. The objective was to facilitate the transfer of information from the suppliers to the manufacturer and vice versa. IT has played a key role in inventory management, determining the specifications of various aspects, and generally throughout the entire manufacturing process. The following section traces the evolution of IT use at Maruti Udyog.

In 1993, Maruti received approximately 130,000 vehicle orders. At the time, no IT infrastructure was available and IT functions were outsourced. In the year 1994–95, mainframe machines were implemented, until the 1995 switch from client server technology to Oracle RDBMS systems. From 1995–96, the entire IT system was redesigned, and an internal mailing system introduced in 1996, the first Indian company to do so. Maruti has ten regional hubs in India including Mumbai, Chennai, and Guwahati, which are all connected through Very Small Aperture Terminals (VSATs) which enhance the speed of communication between the hubs. All vendors and suppliers can connect to the extranets, and secure Internet-based connections are available for the electronic transfer of funds.

In 2001, e-business was launched which further enhanced the concept of true value. Maruti was now able to look beyond the supply and delivery of products and begin recycling and reselling products, as well as offering value-added services. One year later, in 2002, Maruti began offering services such as used car sales insurance and fleet management. Today, all the dealers are connected electronically with the manufacturing unit and payments can also be made electronically by linking dealers and banks, thus enhancing the speed of the transfer of funds. Maruti's overall expenditure on IT amounts to about 0.5% of its sales turnover.

7.2 Information Resources

Maruti Udyog promotes the availability and sharing of accurate information with all its stakeholders, i.e., shareholders, suppliers and vendors, employees, and dealers. The company's management has recognized that computerization is essential for accurate and quick availability of information, and relies on IT to manage information throughout the organization. Access to information available electronically is provided through online menus/folders/queries via the company's local area network (based on a fiber-optic backbone), which covers all departments. Maruti's IT network is regularly reviewed and expanded, and the company has been rated among the top 20 networked companies in Asia. Customers are provided with accurate information about Maruti's products through leaflets, owner manuals, brochures, and displays at workshops. International distributors receive detailed technical specifications of Maruti Udyog's vehicles for printing their own leaflets. The company also provides service manuals, sales guides, and spare parts catalogs to dealers, distributors, and spare parts retailers in order to carry out proper sales and servicing. Policies on sales and after-sales services and other relevant information are communicated to the after-sales network through flyers and bulletins. The regional offices inform dealers about vehicle delivery times, who, in turn, display this information in their showrooms for customers. Daily updates are sent to dealers and suppliers, giving them the opportunity to take decisions quickly based on actual facts and current information. Maruti is connected with most of its domestic dealers (120 at present) and vendors (117) through e-mail. Exact specifications on parts and other material requirements are forwarded to vendors through annual, short-term, and monthly schedules.

7.3 Observations of IT Application at Maruti Udyog

An analysis of Maruti Udyog's manufacturing unit indicates that the entire unit is linked through its IT network, from supply chain to the end consumer. The application of IT in various functional areas has resulted in numerous advantages for Maruti Udyog, including: (1) improvement in manufacturing cycle time, (2) better costs control, (3) real-time supply chain management, (4) faster order processing, (5) real-time demand supply management, (6) improved interaction with customers, and (7) improved financial management due to real-time transfer of funds.

Maruti has increased its value by collaboratively balancing all resources and optimizing the flow of goods, services, and information from the source to the end consumer. IT has also enhanced the value of the automaker's preferred suppliers. Intermediaries are no longer needed as a result of e-commerce, which has led to improved margins. The use of IT has resulted in cumulative savings in procurement costs by making it possible to identify the cheapest supplier, and has increased overall efficiency. Placing online orders reduces transaction time and costs, and minimizes errors in orders and invoicing. The availability of online up-to-date information enables efficient production planning and control, so Maruti Udyog can quickly switch to other models/variants in case of material shortages (which can also be dealt with promptly). The online financial system facilitates the monitoring of cash flows, of the budget and deviation statements, and material costs, and allows timely invoicing and payments, payments to vendors based on confirmed material supply, the monitoring of inventory costs, monthly profits, etc. Maruti Udyog uses its personnel system to keep track of employee absenteeism and suggestions or recommendations made by employees. The employees' monthly salary slips include information on attendance and the recommendations made by the given employee, which is also available to department managers on a quarterly basis to review employee performance.

An analysis of Maruti Udyog's products in comparison to those of their competitors is regularly distributed to all dealers to support vehicle sales. Stock status reports are sent to all dealers on a daily basis to better position them to respond to customer needs. For employees, information ranging from attendance/leave status, recommendations and suggestions made, to reimbursement status is available, which helps resolve many misunderstandings. Such information resources are not only used for information purposes, but have become an important basis for taking policy decisions, determining the effectiveness of actions taken, and for deciding the future course of action. Maruti has also developed a comprehensive Management Information System (MIS) through which information on the company's control parameters is made available to the top management for review in management committee meetings. Divisional and departmental performance is measured in terms of target achievement and depending on the degree of achievement, strategies are modified and adapted to realize the company's goals.

Fig. 5 demonstrates that Internet and e-business technology offers a four-level opportunity spectrum. Simple Internet connectivity represents the first level of opportunity. By installing a public Internet facility, all business community-related communication and coordination activities can be successfully performed. Secondly, intranet for the seamless integration of information and workflow across the various locations of the company can be set up, as well as a global intra-office coordination and communication link for workflow management, using the same Internet technology and thus moving to the second level of opportunity. Third, when intranet is expanded to include suppliers and distributors for seamless sharing of business data, extranet is created, a 24-hour online business channel where ideas, information, work, and capital flow synchronously. Thereby, the third level of opportunity is reached. Finally, when an organization creates a Virtualnet by setting up a virtual organization, interactive marketing facilities can be provided to customers around the clock. The highest level of opportunity can be attained when a company is connected with remote customers to provide them with interactive customer service.

The first level focuses on general opportunities available to a business community. In the second level, the operational efficiency of a multi-location organization is enhanced. The third level concentrates on improving the operation's efficiency across the entire value chain. The final level entails the creation of a closed community involving the company and its customers.

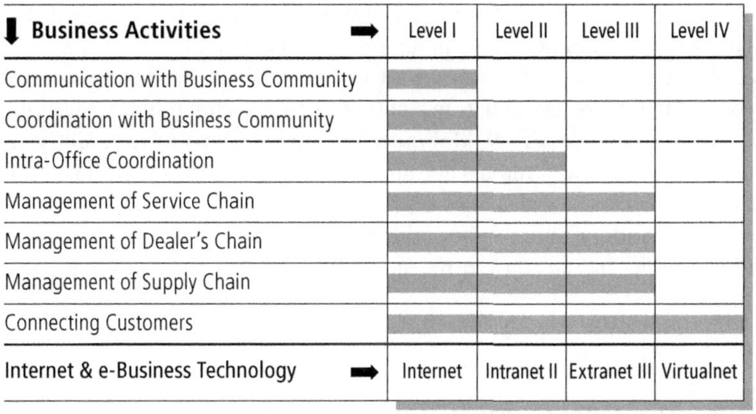

Fig. 5. Opportunity Spectrum of Internet and e-Business Technology

Global automakers like Ford, GM, and Chrysler make use of the Internet's strengths and have created a global network for the collaboration, communication, and the conduct of business transactions with their partners, suppliers, dealers, and customers. They developed Internet-enabled business processes (i.e., e-business processes) and are reforming the automotive sector's production and retail systems. e-Business processes provide efficient and time-saving tools for transmitting information from customers to manufacturers and suppliers. In view of the success of American automotive companies, Indian counterparts are also innovating their production distribution and servicing systems to deliver better products and services to retain their customers and attract new ones.

7.4 Reaching Out to Customers

All major automakers in India are vying for a place in the new net-centric economy. Every vehicle manufacturer has its own Web site which provides information on their cars and a company profile. Most of the car dealers also have or plan to create a Web site. Some focus on getting to know their customers better, and others on improving their supply chains to become more efficient and transparent, or on establishing electronic market places. Maruti's Pune-based ancillary, the Kalyani group, has created a business-to-business portal for automotive parts (i.e., an auto parts exchange), and plans to carry out its business transactions involving approximately $286 million via the Internet (Dubey and Anand, 2000). Daewoo Motors India personalized its servicing strategy—customers register their vehicle on the company's Web site and receive automatic reminders when their vehicle is due for servicing. Complete online transaction processing between Maruti Udyog and its 160 dealers has decreased Maruti Udyog's order processing time to 15 minutes. Tata is connecting its entire network, making it possible for dealers to forward accurate information on the number of vehicles ordered. Nearly all indigenous e-business transactions in the Indian automotive industry are business-to-business transactions. However, business-to-customer transactions in this sector remain negligible. Tata and the automaker Mahindra & Mahindra Limited (M&M) have recently launched an e-business venture, and India's automakers have confirmed their interest in using the Internet for business transactions with suppliers, a step which heralds the dawn of a new era in India's automotive industry. The forces and factors that will shape the outcome of the Indian automotive industry's e-business initiatives have yet to be determined.

7.5 IT and Internet Culture

Maruti Udyog has evolved into an IT and Internet-oriented company, which has become ingrained in its business culture. IT and Internet technologies are used extensively for planning, scheduling, and for monitoring procurement, production, and distribution processes. The automaker spends nearly 0.15% of its turnover on IT (approximately $2.8 million). Maruti Udyog uses Enterprise Resource Planning (ERP) for the procurement and management of workflows, and the Internet is the main means of communication with business and other professional communities, and is used for dealing with enquiries from customers, suppliers, and distributors. Maruti Udyog's 2MB leased line from Tata Communications[1] keeps all key employees connected. The company's home page is quite informative and user-friendly, providing precise information on prices and products, and the possibility for customers to submit queries by e-mail. However, one of the visible drawbacks of Maruti's home page is that it is quite static and is not an interactive, business-enabled Web site.

[1] Formerly known as VSNL, Tata Communications is India's largest telecommunication company.

Maruti Udyog also makes extensive use of intranet technology for the management of workflows and to keep in touch with global business partners and regional sales offices. An integrated intranet-enabled management information system unifies the geographically separated offices. Maruti began using extranet technology for business-to-business online transactions to improve coordination, and to reduce transaction costs and cycle time. Today, 80% of Maruti Udyog's $2 billion business with 160 dealers is carried out via an electronic ordering system. Extranet has improved information flows and keeps the automaker updated on the status of vehicle stocks at every dealer outlet, synchronizes vehicle production and dispatch schedules, and allows the company to determine the production schedules of various car models in accordance with the pattern of demand. Thus, the time and cost of distribution has been reduced and the quality of vehicle delivery services significantly improved. Dealers use extranet to place orders online and to monitor vehicle delivery status. In light of the success of using extranet for transactions with dealers, Maruti Udyog plans to extend similar facilities to include its suppliers, and to create online payment gateways.

Maruti Udyog and seven other major automakers—Hindustan Motors, Bajaj Auto, Ashok Leyland, Hero Group, Mahindra & Mahindra, Tata Engineering, and TVS Udyog—formed an alliance and launched a vertical portal for e-procurement to improve supply chain management and reduce material procurement costs. Maruti's business-to-business (B2B) system is operational, and the company's e-business team is preparing to establish a business-to-consumer system (B2C).

With regard to customer service, Maruti Udyog has launched a pilot Customer Information Center for customers in the Delhi and Gurgaon areas using an outsourced call center partner (GE Capital). Customers can call GE's call center toll free and are connected to Maruti Udyog's Service Provider. This is in line with today's approach to customer relations, which requires personalized customer interactions via new technologies that focus on the needs of customers throughout the product and services lifecycles. Maruti plans to expand this service across India once it has gained sufficient experience from the pilot project. A comparison between Maruti Udyog and Toyota as a competitor is depicted in Table 9.

8 Conclusions and Challenges

Today, automakers' key concerns are cleaner and renewable sources of fuel, more user-friendly cars, recycling, and the reduction of costs. Services have evolved into major contributors to automotive companies' revenue and profitability. Automakers are exploring new value drivers, such as investments in technology and product-related services like finance, insurance, after-sales service, maintenance, and buy back options. India is emerging as an outsourcing hub for automotive components and manufacturing, owing to the country's costs and quality advantages. The case of IT usage at Maruti Udyog clearly demonstrates how IT has successfully been deployed to offer various services requested by customers. The company now offers

Table 9. Comparison of Maruti Udyog and Toyota

Market Share	Maruti Udyog 55.79%	Toyota Kirloskar Motors 20.14%
Common Situation	Global competition; supply potential is twice as high as demand; Internet technology offers opportunities to improve relationships, reduce cycle time, cut costs of distribution and procurement, but requires preparatory work and organizational commitment	
Differentiated Challenges	Retain leadership in economic car production and capture a slice of the new growth in other segments	Find gaps in the market and develop vehicles that meet customers' aspirations and needs
Actors	Production and Marketing Managers	Production and Marketing Managers
Processes	User Internet, intranet, extranet for business communication; workflow management, retail and supply chain management, call center for customer service	Conducting market research to identify opportunities and use proven processes to develop superior vehicles for Indian roads
Learning	Process-focused, improvement of operational efficiency of supply and distribution chains	Product-focused, identification of vehicles that customers want, but which are not yet available
Actions	Introduced extranet for dealers for online ordering, billing, and monitoring of vehicle delivery status	Introduced multi-utility vehicle that can accommodate large families and can be used in all terrains of the country
Performance	Managed to reduce the time and cost of distribution, and to enhance the quality of vehicle delivery services	Product received instant approval by the customers

Source: Majumdar and Gupta (2001)

vehicle financing services, insurance, supply of accessories, buy back options, and refinancing (Fig. 1). IT has also been implemented in all areas of management, including finance, marketing, HR, supply chain management, distribution management, etc. (Fig. 2). Maruti Udyog is one of India's leading investors in IT systems within the passenger vehicle sector. Investments in IT have led to the establishment of a supply chain network geared toward the real-time supply of inventory. However, IT also has its limits. Investments in new designs, the upgrading of standards, and development of alternate fuel technologies require activities that go beyond the capabilities of IT alone. Progress in this area necessitates a regulatory and political mandate and a sincere commitment on the part of industry. Issues such as fuel efficiency, energy conservation, safety, and convenience are all factors that have to be taken seriously to advance the sustainable management of the automotive industry. The competitive advantage of certain features today will not last, but will instead be imitated by competitors. King and Lyytinen (2006) assert that the significance of the "so called modular vehicle platform" will increase. That is, vehicles will consist of an expensive platform which incorporates the major engine- and control-related systems. Interchangeable body components will enable vehicle reconfiguration over time. Customers would lease the platform through a long-term contract and reconfigure the vehicle's parts and components as required. Record systems facilitate such long-term strategies and can assist considerably in the design, manufacture, and operation of such vehicles. Future trends suggest that the ownership model will eventually be replaced by a lease model. If this is the case, fundamental shifts in technology and design may be required to meet global demand.

With regard to regulation at the industry level, India needs to establish a comprehensive regulatory framework making the adoption of IT-driven technology mandatory (Alexis, 2003). A lack of appropriate rules and regulations acts as a deterrent for both manufacturers and consumers to adopt and subsequently benefit from technology. Indian vehicles need to undergo a major shift from the presently mechanical system to an electronic one. Until market forces are mature enough to impel the adoption of technology, regulatory frameworks will have to play a key role. Customers in India need transport solutions that are already available at the global level, which would increase productivity and total cost effectiveness. For example, Sundaram Clayton in India, which developed India's first indigenized antilock brake system (ABS), is advocating the implementation of regulations making ABS standard in at least some vehicle segments, including hazardous goods carriers, long-distance buses, etc. ABS is particularly important in India with its distinct geographical conditions and ever-increasing traffic fatalities. Only the introduction of an explicit regulation would generate interest in such technologies. However, so far the government has remained rather passive in this regard.

Furthermore, Maruti Udyog, India's biggest vehicle manufacturer, ought to embrace virtualnet technology which enables seamless connectivity from suppliers to end users. Customers should be in the position to place orders online, which would reduce the waiting period normally linked to the reservation and delivery of the vehicle to the end consumer. Payment could be transferred directly from the customer's account to the manufacturer's. Enhanced interaction and involvement with customers would further increase the company's value. Though Maruti has made some changes in product design, these have basically been superficial in nature. Changes in product design need to be more genuine, and will require some modifications of current technology. Urgent improvements to vehicle designs are necessary to prevent fatal accidents and to minimize accident-related injuries and damages.

For any business to remain competitive, it is essential for different functional areas to contribute to the business' growth. IT can function as a facilitator, but not as an end in itself. A good blend of different functional areas is crucial. If, for example, the product itself has inherent weaknesses, it is only a matter of time until overall quality suffers. To avoid this pitfall, IT must be seamlessly applicable across different functional areas to produce a more efficient and customer-friendly product. In addition, international pressure on the corporate sector is pushing companies to consider sustainable technologies. Automakers must prioritize certain aspects to remain competitive in the global market. IT cannot substitute fundamental shifts in technology. But it can function as a facilitator, without, however, being an end in itself.

References

ACMA (2006) www.acmainfo.com

Alexis K (2003) Industry seeks tech norms for auto industry growth. http://www.zdnetindia.com

Battelle (2005) Global R&D Report, Battelle, Columbus

Capitaline Plus Database (2006) http://www.capitaline.com

Carli G, Scialdoni R, Tronci M (2006) Life cycle design management in the automobile sector: A framework. www.liedenuniv.nl/interfac/cml/ssp/projects/chainet/abscarli.pdf

Dubey R, Anand M (2000) It's Rush Hour Traffic on the Cyber Highway. Buxine's World, May 22, 2000

Forbes N, Wield D (2000) Managing R&D in technology-followers. Research Policy 29(9):1095–1109

Ford Motor Company (2006) Our route to sustainability: Connecting with society—Ford summary sustainability report 2005/06. Ford Motor Company, Dearborn

Gereffi G (1994) The organization of buyer-driven global commodity chains: How US retailers shape overseas production networks. In: Gereffi G, Korzeniewicz M (eds) Commodity chains and global capitalism, Greenwood Press, Westport

Gibson C F, Nolan R L (1974) Managing the four stages of growth. Harvard Business Review 52(1):76–88

Helper S (1993) An exit voice analysis of supplier relations: The case of the US automobile industry. In: Grabher, G (ed) The embedded firm. Routledge, London

Hobday M (1995) Innovation in East Asia. Edward Elgar, Aldershot

Humphrey J (1995) Industrial reorganization in developing countries: From models to trajectories. World Development 23(1):149–162

Humphrey J (1999) Globalisation and supply chain networks: The auto industry in Brazil and India. In: Gereffi G, Palpacuer F, Parisotto A (eds) Global production and local jobs. International Institute for Labour Studies, http://www.ids.ac.uk/ids/global/pdfs/jhilo.pdf

Kanungo S (1999) Making information technology work. Sage, New Delhi

Kim K (1994) The mechanism of growth of the Korean automobile industry from the technological point of view: The case of Hyundai. Sa-hoi Kyung-Jai Pyung Ron 7 (in Korean)

Kim L (1997) An imitation to innovation: The dynamics of Korea's technological learning. Harvard Business School Press, Boston

King J L, Lyytinen K (2004) Automotive informatics: Information technology and enterprise transformation in the automobile industry. In: Dutton W H, Kahin B, O'Callaghan R, Wycoff A (eds) Transforming enterprise. MIT Press, Cambridge

King J L, Lyytinen K (2006) (eds) Information systems: The state of the field. Wiley, Chichester

Laigle L (1995) De la sous-traitance classique au co-développement. Actes du GERPISA, Université d'Evry-Val d'Essonne No. 14:9–22, cited in Humphrey J (2000)

Assembler-supplier relations in the auto industry: Globalization and national development. Competition & Change 4:245–271
Lee H L, Amaral J (2002) Continuous and sustainable improvement through supply chain performance management. Supply Chain Performance Management, Stanford Global Supply Chain Management Forum, SGSCMF-WI-2002
Lee H L, Whang S (2001) E-business and supply chain integration. Stanford Global Supply Chain Management Forum, SGSCMF-W2-2001
Majumdar S K, Gupta M P (2001) E-business strategy of car industry: SAP-LAP analysis of select case studies. Global Journal of Flexible Systems Management 2(3)
Mcginn D, Kudyba S, Diwan R (2002) Information technology, corporate productivity, and the new economy. Quorum Books, Westport
Nolan R L (1979) Managing the crises in data processing. Harvard Business Review 57(2):115–126
Rosenberg N (1996) Uncertainty and technical change. In: Landau R et al. (eds) The mosaic of economic growth. Stanford University Press, Palo Alto
Wise R, Baumgartner P (1999) Go downstream: The new profit imperative in manufacturing. Harvard Business Review 77(5):133–141
WTO (2003) International trade statistics. WTO, Geneva

Information Technology and Changing Institutional Systems as They Shift from an Industrial to an Information Society: An Initial Comparative Analysis of the USA, Japan, and China

Charla Griffy-Brown, Bing Zhu, and Jianbin Jin

Abstract This study explores the institutional changes emerging in Japan, China, and the USA and, within this context, whether businesses broadly leverage IT to redefine relationships and modify their operations. A qualitative investigation assessing institutional changes in the three countries was completed to determine the constraints of such an institutional comparison. Subsequently, empirical data on small- and medium-sized enterprises was collected to establish whether there was any indication that IT-related institutional changes are taking place outside multinational enterprises.

Key words: Institutional Change, Innovation, Small- and Medium-Sized Enterprises, IT, Institutional Comparison

1 Introduction

There is little question that innovation—the development and commercialization of new technologies—is a significant driving force of sustainable economic growth (Romer, 1986; Grossman and Helpman, 1991). The OECD (2001) confirms that growth depends on building and maintaining an environment that is conducive to innovation and the application of new technologies. Additional studies assert that technology on its own only represents a potential which can be fully exploited when

Charla Griffy-Brown
Associate Professor and Discipline Lead of Information Systems, Pepperdine University, Graziadio School of Business and Management, 24255 Pacific Coast Highway, Malibu, CA 90263-4858, USA, e-mail: charla.griffy-brown@pepperdine.edu

Bing Zhu
Department of Chemical Engineering, Tsinghua University, Beijing, China

Jianbin Jin
School of Journalism and Communication, Tsinghua University, Beijing, China

M. V. Van Geenhuizen et al., (eds). *Technological Innovation Across Nations*,
DOI: 10.1007/978-3-642-00158-1_4,

accompanied by institutional change (OECD, 1997). Furthermore, while technology offers net gains by increasing the efficiency of new goods and processes (Romer, 1994), the pace of institutional change is often much slower than that of technological change (Ng and Griffy-Brown, 1999).

Nevertheless, the global emergence of IT coupled with a new wave of globalization and growing economic interdependence compels societies to transform their institutional structures. Often, such changes are confronted with great resistance due to the perceived or real social and political costs (Ng and Griffy-Brown, 2001; Pempel, 1998). Japan, China, and the USA are witnessing increased IT utilization (Telecommunications Council, 2000). The institutional consequences associated with this shift can be profound. However, if the purpose of introducing IT is limited to the replacement of part of the workforce to improve productivity—as was the case with automation—its full benefits cannot unfold. When technology is simply mapped onto rigid and often futile practices, it may only succeed in increasing the efficiency of a process that has already become superfluous. In this regard, the OECD (1997) analyzed the potential of IT to 'automate' and to 'informate', and concluded that organizations which introduce IT place more relative emphasis on its automate function. However, a firm can become more 'flexible' if it shifts the balance from the 'automate' option toward the increase of transparency and communication (Griffy-Brown et al., 2002; Griffy-Brown and Watanabe, 1999). The firm would acquire a new set of skills, more efficient processes, different product life cycles, and establish a dynamic interaction with its customers. The relations between the firm and its stakeholders would transform as well (Peterson, 1997). Such changes require fundamentally new structures for corporate governance, as well as formal and informal limitations and approaches.

Our research indicates that the 'flexibility' of institutions is important and is evident in the interplay between IT and institutional change in the three nations examined. IT also promotes flexibility in a self-propagating way. While not the only explanation for the trends observed in the USA, Japan, and China, it does provide insight into the role information systems now play within institutional dynamics and in the clash of global versus local economic systems more generally. Such institutional dynamics are particularly relevant for the success of businesses in developed economies. These dynamics also have significant consequences for developing economies, since the dialectic between global and local forces continues to create a turbulent environment for business enterprises worldwide.

Section 2 introduces the theoretical framework of our analysis, while Section 3 describes the process of data collection and methodology. Section 4 addresses our research question and elaborates on the notion of 'institution', thus enabling us to compare the transformation of institutions in Japan, China, and the USA. Section 5 applies empirical data to compare the role of IT in the three countries' institutional transformations. Based on these qualitative and empirical analyses, Section 6 discusses broader implications and draws conclusions.

2 Theoretical Framework

Strategic thinking and planning are essential for firms of all sizes in a globalized economy. The business world faces considerable challenges such as deregulation, new technologies, economic and political pressures, and increasing globalization and competition. These challenges require distinct strategic approaches. Scholars such as Covin and Slevin (1989) consider an entrepreneurial approach essential, since it reflects a firm's propensity to engage in innovation accompanied by a proactive, risk-seeking, and competitive inclination to achieve its strategic objectives (Kyobe, 2004). Stata (1989), Mintzberg (1990), Senge (1990), and Dodgeson (1993) also regard strategy as being not so much about the selection of markets and market positions, but about the building of key internal capabilities that can be sustained. According to these scholars, competitive advantage is rooted in a company's ability to innovate, learn, leverage relationships, and inspire vision. Furthermore, the success of a strategy is often linked to the development of fundamental and distinctive capabilities, which enable a firm to not only enhance its operational efficiency but to also create a sustainable strategic advantage. Information Technology (IT) and Information Systems (IS) play a significant role in promoting this goal. Henderson and Venkatraman (1999), and Sabherwal and King (1991) argue that the effective implementation of IT shapes many organizations' strategies. Other researchers have likewise looked at IT's operational use within different national contexts (Blili and Raymond, 1993; Neidleman, 1979; Delone, 1988; Bridge and Peel, 1999). All of these studies link IT to institutional change, largely through the transformation of relationships (e.g., business-to-business relationships (B2B)) or through operational changes (i.e., using IT to modify operations and thus enhancing operational efficiency). Our study builds on previous research by applying the theoretical context of IT as a self-propagating function with a positive effect based on evolutionary theory (Watanabe et al., 2003). This approach provides insight into IT and institutional transformation in three specific national economic environments, which cannot be directly compared with one another because of distinctive data sets and considerable divergence in institutional structures. The framework illustrated in Fig. 1 is applied to China, Japan, and the USA. Based on the above-mentioned assumptions, this figure depicts the hypothesis that B2B interactions trigger institutional change, as does the implementation of IT to transform internal operations. Both of these factors lead to the modification of business processes and of the way business is conducted. Data revealing that such activities are fundamental elements of an organization's business practice would indicate that institutions are dynamically changing as a result of information technology (Watanabe et al., 2003). Mathematical modeling would, however, be required to explain this causal relationship.

This study aims to establish a framework which facilitates the comparison of the three countries being studied and to determine whether there is any indication that IT in fact plays a role in institutional change, rather than simply assuming that it does and comparing the scope of change based on a potentially erroneous assumption. Therefore, this investigation explores and compares the levels of B2B activity in the three countries, as well as the levels of IT use within SMEs, to verify whether institutional change is occurring.

This study seeks to answer the following research questions:

1. What institutional changes are emerging in Japan, the USA, and China?
2. In this context, are business institutions leveraging IT to redefine relationships and modify their business operations, consequently contributing to institutional transformation?

In order to answer our first research question, a qualitative investigation exploring institutional changes in the three countries was carried out. We collected and evaluated empirical data on the basis of the theoretical framework illustrated in Fig. 1 to answer our second research question.

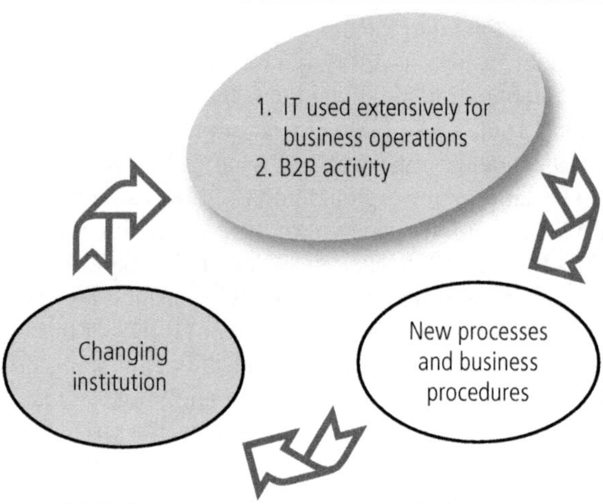

Fig. 1. Proxy Measures for IT and Institutional Transformation

Extensive B2B activities and IT use in SMEs is an indicator for institutional change as a result of the implementation of IT. Further mathematical modeling is required to elucidate the magnitude and scope of this change. However, given the three countries' considerable and dynamic institutional differences, it is important to first establish the nature of the individual institutional systems, and whether there is any indication that IT visibly cultivates change at a fundamental level, prior to undertaking any phenomenological analytical comparison.

As a result of the exploratory nature of our research, the empirical data will not be able to explain the extent of institutional change, but aims instead to identify indicators that such change is indeed taking place. The purpose of our research is to explain the reasons for this change. Phenomenological modeling would be the next step in subsequent research to evaluate the magnitude of change and mathematically describe the relationships involved.

3 Methodology and Data Collection

Our empirical data are derived from several sources. In Japan, meetings were held with Japan's Small Business Research Institute and with company executives in the mobile phone industry, as well as in the IT industry. In addition, data on approximately 16 different SMEs were collected.[1] Primary data on SMEs were compiled from the Management and Coordination Agency and the Establishment and Enterprise Census of Japan. Data were also gathered from the Small and Medium Enterprise Agency and the National Small Business Information Center. Our analysis particularly focused on data based on the Small and Medium Enterprise Agency's 2000 Survey of Creative Corporate Activity, which comprises data from over 800,000 small- and medium-sized enterprises in Japan. Qualitative data were collected through interviews with SME business leaders and owners.

In China, meetings with government officials and university affiliates took place to establish the required contacts for the collection of data on corporations. Data from the China Annual Report of Information Industry, the China Statistics Bureau, and the Annual Report of the Internet in China (CNNIC) were used.

U.S. data were derived from surveys conducted by the Small Business Association, as well as from interviews, focus groups, and related literature. All qualitative data were collected using the triangulation methodology developed by Yin (1994) for qualitative research. This entailed the use of multiple sources of data, such as participant observation, open/structured interviews, and public and private documents to facilitate the mapping of instances that demonstrate the strategic use of information resources by SMEs and of institutional change within all three contexts.

4 A Qualitative Analysis of Institutional Systems in the USA, Japan and China for Comparing the Role of IT

We use the term 'institution' in our study to refer to the "humanly devised constraints that structure human interaction" (North, 1994, p. 360). These constraints consist of formal ones, such as rules and laws, and informal constraints including norms of behavior, conventions, self-imposed codes of conduct, and their enforcement characteristics. Taken together, they delineate the incentive and punitive

[1] For the purposes of this investigation, small- and medium-sized businesses are generally defined as business establishments with under 300 employees, under 100 employees in the wholesale sector, and under 50 employees in the retail and service sectors. Small-scale businesses are defined as business establishments with one to 19 employees (between one and four employees in the wholesale, retail, food service, and service sectors).

structure of societies and economies (North, 1994). We use the term 'institution' in the context of an economic system which consists of multiple institutions at the national, intermediate, and corporate levels (Gao, 2001). This economic system encompasses all multilateral relationships between corporations, citizens, banks, as well as bilateral relationships between trading partners, employers and employees, and producers and consumers (Gao, 2001; Hollingsworth and Boyer, 1997). Furthermore, these relationships are integrated, and nations use them to determine how to organize production and distribute welfare (Gao, 2001). The framework developed below explores the institutional configuration in Japan, the USA, and China, highlighting key differences in their flexibility and the role of IT in this context. Tables 1 and 2 outline the two primary areas of discussion analyzed for all three countries. Table 1 presents the key qualitative institutional indicators identified. The key IT indicator used for means of comparison is the role and the use of the Internet in business practice in all three countries, particularly in SMEs. Table 2 compares general Internet usage to set the overall context explored in our analysis, namely Internet usage in business.

Table 1. Key Institutional Indicators

Indicators	China	Japan	USA
Relationships among firms	Changing ownership struture, not IT-related	Based on historical relations	Dynamic and largely IT-enabled
Process and decision-making	In flux	Consensus and not IT-driven	Driven by new decision-making information and support systems (ESS, DSS, MIS)
Transparent financial system	Not transparent or IT-enabled	Increasing transparency and IT-enabled	Transparent and highly regulated IT accountability

Building on our theoretical framework, Section 5 takes a closer look at IT use and penetration in all three countries to determine the occurrence of institutional change.

4.1 Institutional Change in Japan

During Japan's 'catch-up' period starting in the 1950s and 60s, manufacturing technology was considered crucial for sustaining industrialization. Japanese institutional systems were accustomed to this technology. Japanese business management systems, such as lifetime employment, the seniority system, lean production (e.g. TQC,

Table 2. Key IT Indicator Comparison: Internet Usage

	Population (est. 2006)	Internet Users in Dec. 2005	% Population Penetration	Increase in Use 2000–2005
USA	299,093,237	203,824,428	68%	113.8%
China	1,306,724,067	111,000,000	8.5%	393.3%
Japan	128,389,000	86,050,000	67.2%	83.3%

Notes:

1. The Internet statistics are derived from Internet World Stats December 31, 2005.
2. The demographic population figures are based on data published in World Gazetteer, updated on January 1, 2006.
3. The usage numbers are from the International Telecommunications Union.
4. Figures on the increase in use are derived from Internet World Statistics, December 31, 2005.

JIT, and Kaizen), the main banking system, and *keiretsu*, facilitated the adaptation of technology to manufacturing firms' institutional systems. Thus, the highly efficient, closed network among related entities thrived and leveraged technology for their own strategic advantage (Watanabe, 2002; Katz, 1998; Gao, 2001).

In the lifetime employment and seniority systems, for example, Japanese firms invested in employee training to improve their staff's company-specific skills. The *keiretsu* system, a long-term oriented relationship among Japanese firms, was rooted in the shortage of resources following World War II. By establishing closed business networks among related firms, businesses were able to efficiently coordinate the production and distribution of goods based on technical and institutional stock restricted to these rigid and related business entities. Technology and knowledge were often not transferable beyond the scope of a corporation or industry's boundaries because they were unique to each 'family'. Manufacturing technology enhanced the coordination, distribution, and production processes within this closed system.

Japanese manufacturers developed their products based on their own in-house technology. To warrant the quality demanded by customers of their products and accessories, firms also favored in-house procurement of manufacturing parts, or relied on their *keiretsu* partners to deliver the required parts. In other words, Japanese firms used an 'individual language' that excluded any business that was outside the family. This mutually dependent system worked effectively to distribute risks among related firms by variably distributing the losses and profits, independent of each deal.

Japan's traditional business systems also established separate, closed networks with the main banking system, which provided banks with inside information on their partner firms and the possibility to monitor their financial transactions. Within these networks, each system functioned complementarily and dependently, achieving high productivity in this protected environment with its specialized line workers and stable supply of products. Implicit procedural rules among related entities and specific communication 'languages' within these closed networks were developed, which excluded anyone located outside of these tight-knit 'families'.

During the 1970s, the 'catch-up' economy and the associated technologies matured, along with the processes they enhanced. Two significant shifts realigned Japan's institutional configuration and its supporting technology. The first was the global shift from trade and production to financial and monetary activity (Gao, 2001). The second was the shift from social protection to the release of market forces (Gao, 2001). The Japanese institutional configurations no longer corresponded with market pressures following these shifts. Additionally, the technology that had been responsible for making these institutional processes run so smoothly no longer created an advantage in an institutional system that had become unsuitable. Not surprisingly, the Japanese bureaucracy, though it had also matured, resisted change.

In a fully integrated economy, global market and political forces drive domestic institutional change. An integrated economy evolves once a domestic economy and its large firms reach maturity and must compete globally in order to survive. However, in the case of Japan, the market-conforming policies introduced in the previous era had been built on state policies designed to isolate and preserve existing industries and institutional structures. Consequently, institutions had no incentive to change (Katz, 1998). Although these strategies had in the meantime evolved into market-defying policies, until the mid-1990s, Japanese institutions vehemently opposed the transformation of their institutional systems to adapt to globally competitive structures (Gao, 2001; Katz, 1998). However, the technology underpinning the country's previous tremendous economic growth no longer created any advantage for Japanese businesses.

This is where information technology—rather than technology that automates manufacturing—comes in to play a critical role. An institutional transformation requires bureaucracy to become transparent in terms of process and decision-making. Institutions become more accountable to the elected representatives and the public. Though this transformation is beginning to take shape at the political level, and since regime shifts have already been analyzed (Pempel, 1998), the institutions that are of particular interest in the context of our study are firms.

At the firm level, corporate governance changes precede process changes, and the key to strategic success in leveraging IT is to develop it for the purpose of enhancing processes. Thus, IT's objective should be the improvement of transparency and communication. As a result of global pressures, such changes were initiated in Japan in the early 1990s. In response to decreasing competitiveness, Japanese companies were diversified, which resulted in changes in ownership. For example, Canon and Sony's portfolios now disclose a 60% foreign stock ownership or foreign institutional investors. The real institutional transformation, however, began to take place with the increasing pressure of the capital markets after 1997. Due to substantial bank failures, the primary financing mechanism no longer came from borrowing, but from issuing bonds. This meant that only firms with a high credit rating benefited from a low capital cost, and the difference between a AAA rating and BBB rating rose from 1% to 4%. Companies realized that they would not survive without institutional reform because they could not afford capital.

Consequently, many firms modified their corporate governance structure to attract foreign investors and increase their credit rating. Some firms implemented a reform of their board of directors, referred to as *shiyoku ginsei*, which stipulated that the board had to include representatives of external shareholders. Other firms introduced stock options. No Japanese firm, however, was flexible enough to introduce both the monitoring system (*shiyoku ginsei*) and the incentive system (stock options).

Japanese firms introduced the multi-dimensional structure along the lines of the U.S. model, granting the business unit or profit center independence (*companysei*). This decentralization of the firm coincided with the centralization of financial markets. However, for these governance and process changes, particularly in terms of transparency and accounting, to evolve, the information system had to be modified. Accounting structures changed with each division being required to not only provide a profit and loss statement, but a balance sheet as well. This reinforced the autonomy and decision-making authority of the division manager.

IT, which addresses the 'communication' aspect of these processes, must be compatible with the new institutional configuration. Thus, as institutions become more flexible through organizational transformations, new information systems are adopted that enhance flexibility. This flexibility is then propagated and improved, thereby creating greater flexibility and increasing momentum in terms of process efficiency. The conceptual result is that flexibility turns into a self-propagating function, as positive feedback creates momentum. Though such developments can be observed in several large Japanese firms, the critical issue is whether or not institutional change is taking place in small- and medium-sized firms as well, which make up the bulk of the Japanese economy. This will be empirically examined in Section 5.

4.2 *Institutional Change in the USA*

As MacRae (1995) argues, America's melting pot generates a nation of new ideas, cultivates a "frontier spirit", and enhances flexibility that fosters the incorporation of IT. Indeed, the country's indigenous qualities effectively accelerated structural change of the U.S. industry in the 1990s (Bisin and Verdier, 2000). While Japan was struggling with the shift from trade and production to financial and monetary activity, as well as from social protection to the release of monetary forces, the USA had undergone this institutional adjustment and, therefore, already conformed to these changes. That is, corporate governance was transparent and deregulation required businesses to engage in fair and competitive practices.

In the late 1980s, U.S. firms began to invest heavily in business sectors in which they had competitive advantages, focusing on core competence management to improve their profitability or return on investment, shareholders' main concern. Furthermore, they began outsourcing certain business operations, such as data processing, and invested in the development of business systems to enhance efficiency. This led to the expansion of the outsourcing market and the mobility of labor in the USA. In contrast to the closed network or family system, this also meant that businesses had to standardize communication protocols and formats not just vertically, but horizontally as well.

The dramatic advancement of IT in the 1990s further propelled the structural change of U.S. industry. With the expansion of network technology, business transaction costs between firms significantly decreased, facilitating the division of business processes into modules for efficient outsourcing. Firms emerged that built on their core competences and competed against each other to reach deals with former partners and rivals.

IT's rapid development, paired with network externalities, enabled quick dissemination of IT-related products and services, and shortened product life cycles which, however, made it more difficult to forecast future product trends. An effective business strategy to respond to the surge of IT, which requires speedy R&D and the timely introduction of products and services into the market to successfully establish a competitive global position, was the establishment of flexible business alliances between firms. Aware of this trend, U.S. firms willingly disclosed business information to find appropriate business partners, resulting in the enhancement of their value chain as a whole.

Finally, increased electronic connectivity evened the asymmetry of information on products between suppliers and customers. Customers could now more easily obtain information on market products through the Internet. Accordingly, suppliers must identify customers' needs in order to be successful in an extremely competitive market.

Furthermore, laws now required firms to become transparent and prohibited conflicts of interest and collusion. To respond to these changes in their external environment, U.S. firms adapted their strategic management to become more flexible by disclosing business information to share their resources with potential partners. This induced the establishment of networks open to any firm, thus contributing to all participants' value added, including customers'.

These structural changes in U.S. firms, which ultimately led to the nation's current success as an information society, were endorsed by firms' efforts to standardize their communication interface, or to establish a 'common language'. In fact, efforts to create a 'common language' posed little challenge for U.S. firms, considering the country's melting pot origin: the USA's heterogeneous environment promoted the establishment of 'standard languages', facilitating direct communication between people and organizations and the formation of open networks with standardized interfaces. Thereby, the acceptance of newcomers and changes becomes much easier. Though there is still room for improvement, the evolving institutional configurations correspond to the emergence of information technology, which has further enhanced U.S. firms' ability to compete in the global environment. Assuming that this qualitative assessment is accurate, the question then is whether IT-related institutional changes are observable, even among small- and medium-sized enterprises which compose the bulk of the U.S. economy. This question will be explored in Section 5.

4.3 Institutional Change in China

Since the onset of economic reform in 1978, the Chinese economy has grown at an average of 10% annually, and even averaged 12% per year during the first half of the 1990s (National Bureau of Statistics, 1992a–1999a, 1993b–1996b, 1994a). This growth has taken place amid diverse institutional structures, such as changing types of firm ownership. Ownership is recognized as being a fundamental institutional arrangement in a socio-economy because it incorporates different incentive structures (Li et al., 2004). In this context, there are three different types of Chinese enterprises distinguished by their coordination of transactions and property rights: (1) state-owned enterprises (SOEs), (2) collectively owned enterprises (COEs), and (3) privately owned enterprises (POEs).

SOEs are owned and operated by the state. The 'state' is not a single entity but consists of several levels. Institutions are subject to diverse institutional arrangements determined by the degree of state involvement (e.g., national or local) and how many enterprises the particular level of government 'owns'. COEs are owned by the employees working for them, but the precise status of property rights is somewhat ambiguous (Li et al., 2004). COEs emerged in the 1950s and employed millions of people that SOEs simply could not absorb. They were considered less prestigious than SOEs at the time and initially performed poorly, in part because they did not enjoy the same allocation of resources, such as bank loans and subsidies, despite being subjected to the same regulations as SOEs. Eventually, COEs were conceded to local governments by the state for financial reasons (Li et al., 2004). POEs include all enterprises that are neither SOEs nor COEs, have relatively clearly defined property rights, and are more market-driven. This category includes privately owned companies, foreign-invested firms, joint ventures, limited companies, and joint stock companies (Li et al., 2004). POEs did not exist from 1949 to 1976 (NBS, 1995a) and reemerged during the 1970s as part of China's economic reform. Even though POEs are not state-owned, they are subject to state interference through an institutional arrangement referred to as *lishu* relationships (Li et al., 2004).

To understand the concept of *lishu* relationships, the different levels and configuration of government institutions in China must be elucidated. China has 31 provinces and municipalities, 334 prefectures (*diqu*), 2,143 counties (*xiang*), 47,806 towns (*zhen*) and city (*qu*) districts. This structure builds on five levels of government: the central government, provincial governments, and regional, county, and town governments. All enterprises are subject to the supervision or *lishu* of one of these five levels of government (Li et al., 2004). The government is in fact required by law to interfere in firm activities through *lishu* relationships. This specifically means that government involvement in POEs through *lishu* includes the naming of firms, regulating their structures (appointments of directors, chairmen of the board, general managers, etc.), reviewing their feasibility studies and business plans, approving their licenses, determining the amount of taxes owed, approving major projects, issuing bank loans, monitoring bank transactions, and even making major operational decisions (e.g., on profit distribution and investment) (Li et al., 2004).

The Chinese government initiated economic reforms in 1978, initially targeted at making SOEs and COEs more efficient. In order to increase efficiency, the central government decentralized authority along two axes: (1) decentralization from central to local governments, and (2) decentralization from the government to economic units (enterprises and households) (Boisot and Child, 1996; Qian and Weingast, 1997; Li et al., 2004). Under this reform, the higher levels of government still controlled the majority of firms, but fiscal responsibility and property rights were defined more accurately. The fiscal reform of 1994 divided the tax revenues into central, local, and communal taxes. Local governments thus began to shift their focus from leveraging relationships at higher levels of government to increasing local tax revenues. This reform also meant that the bulk of central government revenue would now derive from national taxes, customs duties, and consumption taxes, thus significantly reducing revenue once gained from firm profits. That is, while the central government's revenue rests primarily on taxes, local governments' revenues are contingent on market conditions and they subsequently place more emphasis on firm performance than the central government. It must also be noted, however, that lower levels of government deal with a smaller number of firms than the central government. Walder (1995) argues that "local governments with smaller industrial bases have clearer financial incentives and constraints, fewer non-financial interests in enterprises, and a greater capacity to monitor them" (p. 263).

In sum, the most significant institutional changes in China include the rise of POEs and the decentralization from central to local governments. The role of information systems in institutional change in this relatively complex context requires a closer look at how relationships were redefined and whether any operational changes are evident.

5 Comparison of Information Technology Use in the USA, Japan, and China: B2B and IT Use in Business Operations as an Indicator of Institutional Change

5.1 IT and Institutional Change in the USA

The institutional systems in the USA, Japan, and China will be empirically evaluated on the basis of our qualitative framework. Data reveal that SMEs and large enterprises (LEs) in the USA widely adopted the tools of the information economy during the 1990s, thus enhancing their efficiency.

Market research confirms that computer penetration and use in the USA is not only high in large firms but in SMEs as well. Dun and Bradstreet found that by May 2003, nearly 99% of small firms (with 25 or fewer employees) in the USA worked with computers and that 98% of such firms had Internet access (US Bureau of Census, 2004). Surveys on Internet access at manufacturing plants provide similar results (Table 3). 83.9% of all manufacturing plants in the USA use the Internet for business processes, though larger manufacturing plants (with 20 or more employees) are more likely to have Internet access (US Bureau of Census, 2004). Notably, these 'electronic activities' take place throughout the U.S. economy. In fact, manufacturing in the USA has a long history of leveraging information technology to achieve greater flexibility, with some firms having used Electronic Data Interchange (EDI) since the 1960s. The majority of component industries, even in sectors with a previously low use of e-commerce, carry out some transactions online (US Bureau of Census, 2004).

Table 3. Internet Access of Manufacturing Plants, 2003 (in Percent)

Employment Size	With Internet Access
1–4	41.7
5–9	52.1
10–19	64.7
20–49	76.2
50–99	84.9
100–249	91.5
250–499	94.1
500+	94.9
Total	83.9

Source: US Bureau of Census, E-stats, 2004

U.S. firms are increasingly using information technology tools for processes beyond buying and selling, which creates greater institutional flexibility and a greater ability to lead and respond to the marketplace. Firms now use the Internet, for example, for marketing purposes, customer service, finance, accounting, sales force automation and human resources, as well as for supply-chain management and sales. Looking at computer use in more detail, widespread PC use is common among SMEs and LEs in the USA, despite the asymmetry in different sectors (Table 4).

In terms of manufacturing, the levels of orders placed online in the USA are particularly high, namely over 70% across the manufacturing sector for B2B transactions (Fig. 2).

5.2 IT and Institutional Change in Japan

Institutional history in Japan differs, but is gradually developing in the direction of the USA. In order to understand the dynamics of IT more clearly, a look at SMEs reveals a great deal. SMEs, which represent 95% of businesses in Japan, demonstrate broad institutional change. In this context, it is useful to identify the operations and processes for which firms currently use information technology. SME involvement in the business-to-business (B2B) market is of particular interest, since SMEs in Japan play a critical role in many complicated business clusters, often involving

Table 4. Employed Individuals Who Regularly Use a Personal Computer at Work by Industry and Size of the Employer, 2003

Industry	Regularly uses a PC (in percent)	Regularly uses a PC at work (in percent)
Agriculture	31.0	13.2
Less than 25	35.2	14.5
25–99	28.1	13.1
100 and over	15.2	7.7
Construction	31.4	18.0
Less than 25	27.5	14.7
25–99	32.3	18.4
100 and over	36	22
Durable Manufacturing	50.6	38.9
Less than 25	44.4	28.9
25–99	44	29.6
100 and over	52.7	41.8
Non-durable Manufacturing	46.2	35.3
Less than 25	39.8	29.2
25–99	39.1	29.1
100 and over	48.4	37.3
Transportation	47.9	32.3
Less than 25	44.4	27.7
25–99	44.6	31.8
100 and over	49.2	33.3
Communications	75.9	68
Less than 25	61.1	48.4
25–99	63.9	51.7
100 and over	78.5	71.3
Public Utilities	60	49.8
Less than 25	62.7	41.3
25–99	51.6	42.9
100 and over	60.9	51.6
Wholesale Trade	57.6	44.9
Less than 25	52.5	39.3
25–99	52.7	41.1
100 and over	61.8	48.9

Source: US Bureau of Census, E-stats, 2004

relationships with large organizations. The size of the B2B market in Japan is surprisingly small, given the significance of these clusters (Fig. 3). This difference in size between the USA and Japan can partially be explained by consumer emphasis on Internet technologies, as well as the traditional face-to-face business practices which still prevail in Japanese business relationships. Furthermore, while there have been significant advances in IT throughout Japan, particularly in mobile communications, SMEs are, on the whole, investing less in information systems than large

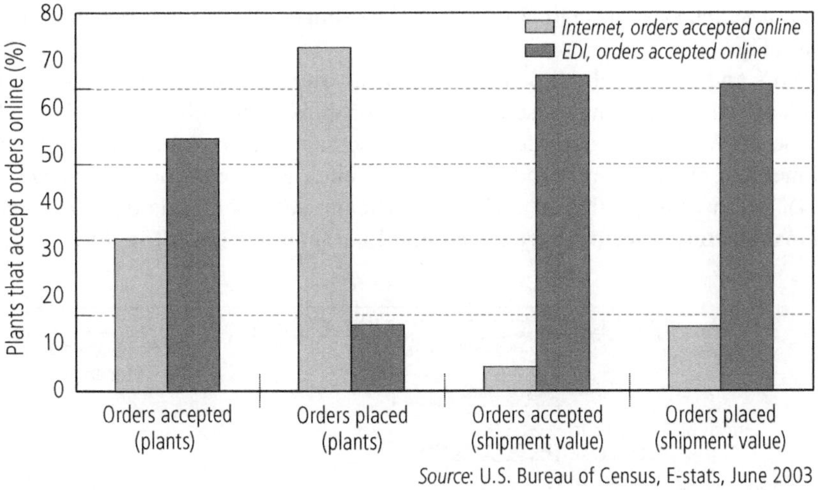

Fig. 2. Manufacturing Plants' Use of Internet and EDI to Place and Accept Orders

enterprises (JSBRI, 2004). In fact, approximately one in ten firms in the small business sector does not invest in an information system (IS) at all.

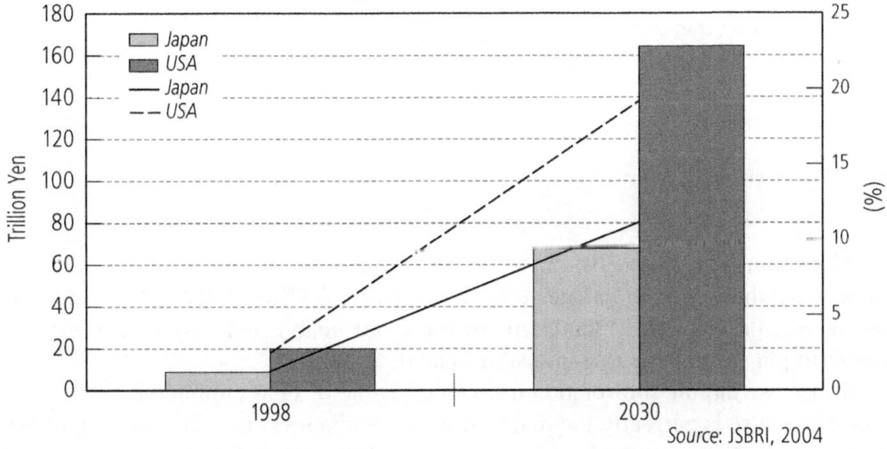

Fig. 3. B2B E-Commerce Markets in the USA and Japan
Note: *The dotted and full lines show previously predicted growth*

When this issue is probed, it becomes evident that the underlying concerns about whether or not to invest in IS are multifaceted. Many of the problems identified relate to human resources (e.g., finding staff who can use the technology or specialists to assist once the IS has been installed), as well as to finding the resources to purchase the systems, and how such an investment affects other potential investments (Fig. 4).

SMEs and even small home offices in Japan are increasingly using the Internet to enhance their customer base and/or relationships within clusters. In fact, while the use of Web sites varies depending on sector and size, over half of all small and medium manufacturers and service companies now have a home page (JSBRI, 2004). A closer inspection of SMEs in particular sectors reveals that the services and manufacturing sectors have the highest Internet usage rates (Fig. 5).

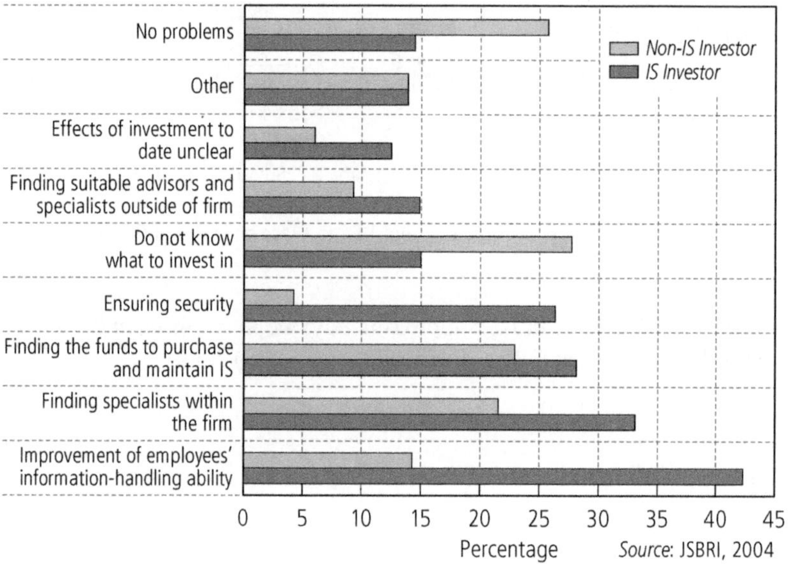

Fig. 4. SME IT-Related Challenges for Investors and Non-Investors in IS

Approximately 20 to 30% of SMEs of all sizes use the Internet in their B2B networks, whose growth in sales tends to be larger (JSBRI, 2004). The new Internet service provider (ISP) landscape featuring affordable and easy-to-use services seems to play a valuable role in institutional development.

SMEs use mobile information devices for some of their communication needs. Cell phone use is currently the main means of communication (Fig. 6). E-mail use follows as a distant second. SMEs are successfully reforming their business and logistics operations by making internal information available online so that it can be accessed from outside the office with the use of mobile devices (Fig. 6).

The data presented so far indicate the spread of IT and its use in B2B, as well as within the context of Japanese business operations. The preceding analysis shows that Japanese firms widely use the Internet to enhance their operational and strategic relationships. That is, certain IT-related institutional changes are indeed taking place.

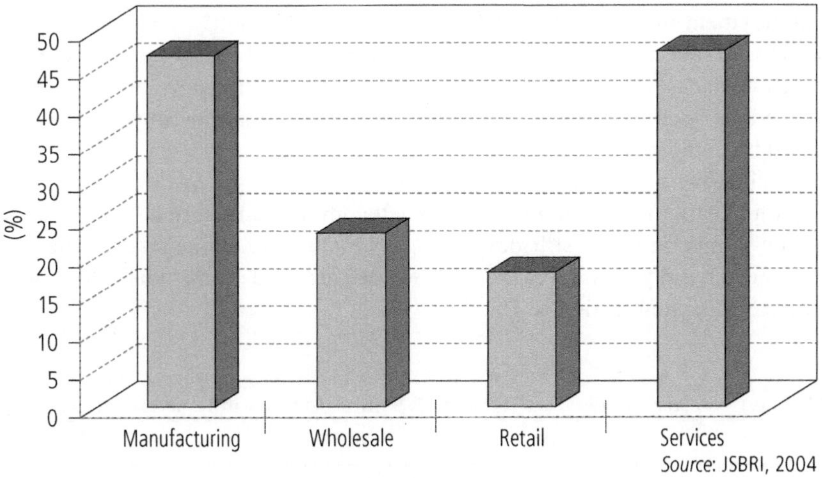

Fig. 5. Percentage of SMEs with a Homepage in 2003

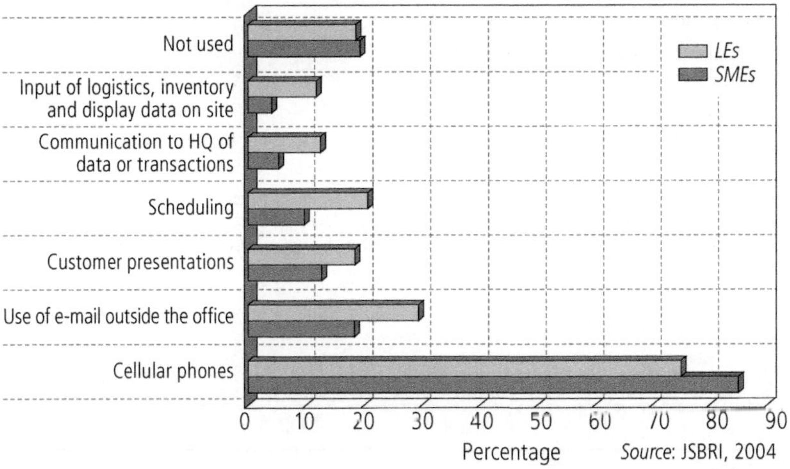

Fig. 6. Use of Mobile Devices

5.3 IT and Institutional Change in China

By the end of June 2004, China had over 3.2 million small- and medium-sized businesses (COMTEX, 2004). According to recent statistics from the China Association of Enterprises, 23 million private and individual firms are involved in industrial production and commercial operations (COMTEX, 2004). The data also reveal that these businesses make up 99% of China's total number of firms. SMEs' end products and service value account for 55.6% of China's GDP. SMEs play a critical role in economic activity, creating job opportunities for an estimated three-fourths of employment in urban areas (COMTEX, 2004). Currently, 65% of patent technologies are owned by SMEs; 75% of technical innovation projects are completed by such businesses, and 80% of new products are developed by them (COMTEX, 2004). Since such high percentages confirm SMEs' economic impact, their use of IT must be explored.

In China, the use of IT by an enterprise, particularly by SMEs, depends particularly on existing infrastructure. China's telecom infrastructure is large in terms of absolute numbers, but is still developing. Nonetheless, China's infrastructure has experienced rapid growth over the past two decades and seems poised to grow even further, as indicated in Tables 5 and 6.

Table 5. A Comparison of East Asia & Pacific Countries and China in Terms of Three IT Indicators

	2000		2003	
	East Asia & Pacific Area	China	East Asia & Pacific Area	China
Fixed lines and mobile telephones (per 1000 population)	148.1	177.6	356.7	423.8
Average cost of local telephone call (US $ per three minutes)	0.0	N/A	0.0	0.0
Personal computers (per 1000 population)	16.1	15.9	N/A	N/A
Internet users (per 1000 population)	24.4	17.4	68.2	63.2

Source: World Bank Group's Web site. Retrieved from http://web.worldbank.org/stats.html

Table 6. The Development of Telecom in China Since 2000

	2000	2001	2002	2003	2004
Mobile phone (millions)	85.26	145.222	206.616	268.693	334.824
Telephone (millions)	108.807	144.407	179.034	263.305	312.443

Source: Based on the Web site of the Ministry of Information Industry, PRC, http://www.mii.gov.cn

Many researchers believe the IT industry has been driving China's economic growth since the 1980s (Tan and Ouyang, 2002). The combined revenues from hardware, software, and IT services accounted for 1.13% of the country's GDP in 1999 (Tan and Ouyang, 2002). In 2005, there were 120 million personal computers in China. The rate of increase in personal computer ownership is 35% per year and, according to CNNIC's latest report, the number of PCs with Internet connection reached 45.6 million in 2005.

Even with obstacles such as China's relatively late entry into the field of telecommunications, Internet infrastructure is growing. Table 7 illustrates this rapid increase. Survey results published by CNNIC in 2005 present the total number of Internet users, computers with Internet connection, international gateway bandwidth, and domain name registration from 1997 to 2001. The number of Internet users rose from 600,000 in 1997 to 103 million in June 2005, and computers with Internet connection increased from 340,000 in 1997 to 45.6 million in June 2005. Total bandwidth increased from 30 Mbps in 1997 to 82617 Mbps in June 2005.

Table 7. China's Internet Infrastructure Growth Since 2002

	2002 (Dec.)	2003 (Dec)	2004 (Dec.)	2005 (June)
Internet users (in millions)	59.1	79.5	94.0	103.0
Computers with Internet connection (in millions)	20.83	30.89	41.60	45.60
International gateway bandwidth (Mbps)	9380	27216	74429	82617

Source: http://www.cnnic.net.cn

In 2002, the majority of China's Internet users (63%) used dial-up services to access the Internet, and 20% used broadband technologies (Tan and Ouyang, 2002). These figures are changing considerably, as indicated in the tables above. Furthermore, according to the latest CNNIC report on China's Internet development, users of broadband now exceed those who use dial-up services to access the Internet.

Internet users in China are also changing in demographic terms (Tan and Clark, 2000; Tan and Ouyang, 2002). Female users now account for approximately 40% of all Internet users, a ratio that has generally remained constant since June 2001 (CNNIC, 2005). While Internet use among individuals aged 35 and above is gradually increasing, 81.3% of users are still under the age of 35 (CNNIC, 2005). IT infrastructure is still inadequate, particularly for small- and medium-sized enterprises. However, this situation is also changing as a result of the Chinese government's effort to reduce the domestic digital divide. Large enterprises' IT infrastructure has improved significantly due to the increase in resources invested in this area. Based on survey data gathered by China's State Economic and Trade Commission, the following figures represent the 100 corporations with the highest levels of IT investment in 2000:

- 16.7% of Chinese enterprises with high levels of IT investment had fully implemented Office Automation and Management Information Systems, while 81.1% had partially employed such technologies.
- 16.5% of the Chinese businesses surveyed actively applied ERP, MRPII, and CIMS, with 50.5% of the businesses partially using these systems, and 33% not using these technologies at all.
- 87% of the enterprises had a company-wide Intranet and Web site.
- 37% conducted some form of e-commerce.

Given that these figures represent organizations with the highest levels of IT investment, they reveal that the infrastructure of large businesses is quite inadequate, and that SMEs' infrastructure is of even poorer quality. A comparison of China's Internet hosting, Internet usage, and access costs at peak and off-peak times clearly shows that China lags far behind many of its Asian–Pacific counterparts in terms of Internet infrastructure availability and usage (Table 8).

Table 8. Critical Infrastructure in Asia–Pacific Nations

Countries	Internet hosts per 1000 population (2004)	Internet users per 1000 population (2004)	PCs per 1000 population (2004)
Australia	197.83	652.8	689.0
China	0.124	71.6	40.3
Hong Kong	110.10	489.1	588.5
India	0.133	32.4	12.1
Indonesia	0.501	65.2	13.6
Japan	128.68	502.0	541.5
South Korea	113.01	656.8	544.9
Malaysia	5.281	386.2	191.6
New Zealand	150.49	819.5	492.7
Philippines	0.791	53.2	44.6
Singapore	116.59	561.2	622.0
Taiwan	138.97	538.1	527.8
Thailand	5.813	112.5	60.0

Source: ITU's free statistics (2004). Retrieved from http://www.itu.int/ITU-D/ict/statistics/at_glance/Internet04.pdf

A number of serious Internet-related concerns have been reported by online buyers (Table 9). These include security, payment inconvenience, product quality, service, and credibility. Such apprehension represents a significant hurdle for both businesses and consumers with regard to increasing the use of Web applications. One of the most significant drivers of automated B2B transactions is the Chinese government, motivated by economic and political reasons, and which will inevitably impact the use of IT by businesses and consumers. Large enterprises also seek increased connectivity to the global economy, as well as improved efficiency among their suppliers, and are therefore, like the Chinese government, advocating the spread of automated B2B transactions.

The most serious impediment to the growth of B2B transactions is the lack of trust among potential participants. Distrust is largely attributable to the fact that China does not have a sound legal and technical foundation which monitors and enforces relationships of trust among businesses. This lack of trust severely affects contract enforcement and free exchange which are defining features of B2B transactions. Furthermore, the current structure of Chinese businesses places middle managers (who are key players in state-owned enterprises) at risk of losing their privileged positions. This factor, coupled with such managers' (usually age-related) lack

Table 9. Major Concerns for Online Buyers

	July 1990	January 2000	July 2000	January 2001	July 2001	January 2002	July 2003	July 2004	July 2005
Security	30%	36.50%	31.80%	31.20%	33.4%	31.00%	25.1%	28.9%	26.9%
Payment Inconvenience	22%	17.70%	13.30%	12.60%	11.5%	11.80%	9.9%	6.8%	6.3%
Product Quality, Service, Credibility	34%	27.70%	28.30%	32.00%	33%	30.20%	40.0%	43.8%	48.4%
Delivery Delay	6%	9.20%	10.10%	9.90%	8.7%	13.90%	7.1%	6.5%	4.9%
Unattractive Pricing	8%	7.80%	7.70%	7.40%	6.6%	6.3%	10.3%	6.0%	5.3%
Reliability On-line Information	N/A	N/A	7.30%	5.90%	6%	6.30%	7.0%	7.3%	7.7%
Others	N/A	1.10%	1.40%	1.00%	.8%	0.50%	0.6%	0.7%	0.5%

Source: CNNIC, 2005 and Tan and Ouyang, 2002

of IT skills, creates an environment in which managers openly resist any shift toward an increased use of Web applications and integrated systems in general. While there are indications that IT-related institutional changes are taking place in the field of B2B, they are difficult to pinpoint given the varied institutional structures involved (COEs, POEs, and SOEs), as well as the different levels of government intervention. The velocity and magnitude of these changes are unclear, particularly given the unusual institutional configuration and its evolving nature. Further empirical work is required to fully dissect the relationship between IT and institutional change in a transitional economy such as China, particularly, given its scale.

6 Conclusion: Summary Comparison of Institutional Systems in Japan, the USA, and China in Effectively Leveraging IT

Our research reveals that U.S. institutions were able to adapt to technology that 'informates' because of the already existing business structure. The data emphasized U.S. businesses' long history of IT-related use and its development. What may appear like a rapid transformation may actually be attributable to a long history of institutional and IT 'mapping'. The accelerated speed of change could also be linked to legislation introduced following the corporate scandals in the late 1990s, thus modifying corporate governance and institutional processes. That is, IT-related institutional changes seem to partially rest on the corporate governance system and firms' ability to create flexible business relationships with other institutional actors.

This flexibility is further enhanced by the U.S. firms' financial structure, which requires transparency. In short, the U.S. system is built on an economic system composed of 'strangers' or 'outsiders', so the rules are explicit and well-defined, which constrain the otherwise fluid institutional configuration. Information technology easily fits into this institutional arrangement, complementing and enhancing communication which, in turn, promotes greater flexibility. Japan's economic system, on the other hand, was set up for 'families', for actors who share a tacit understanding with those they interact with. As a result, the institutional system is very stable but also much less flexible. There is very little transparency since the system's rules are implicit. Until recently, the financial structure was stable yet inefficient and disconnected from global institutional structures. Therefore, information technology did not serve the purpose of 'informating', since explicit institutional rules and open communication are prerequisites.

A comparison of the IT use in business practice revealed a significant divergence between the USA and Japan. The use of IT tools in various sectors by LEs, as well as SMEs, is widespread in the USA, denoting an IT-driven economy. Japan's use of IT for business processes is spreading to SMEs but at a different pace. However, in both cases data indicated that business institutions were leveraging IT to recast relationships and to adapt their business operations, thus contributing to institutional transformation.

The case of China was very difficult to compare. As indicated in the qualitative analysis, the nature of institutional change is not only distinct but occurs on a different scale. There were, however, some indications of a vibrant and growing B2B market, indicating the recasting of relationships to some degree and, subsequently, the occurrence of some IT-related institutional change. However, given the infrastructural constraints and the rapidly evolving nature of China's socio-economic institutional system, it is difficult to clearly identify the individual drivers of this change.

These insights are a very useful foundation for future comparative research in this area. Given the results of our study, it is probably more useful to develop models comparing Japan and the USA and to establish a different model to examine the role of IT and institutional change in China. Such a model could perhaps be used in a comparative analysis between transitional economies, such as India, where the scale and magnitude of change might be similar to that of China. However, for this the qualitative and quantitative framework would first need to be established. Our analysis also reveals that further studies, particularly on modeling, should focus on the rate of IT incorporation into general business practice and the magnitude of business change. Understanding these two factors in a phenomenological sense would provide significant insights into the ongoing relationships between information systems and institutional transformation.

As more firms begin to respond to global market pressures, additional institutional changes may evolve. In a hyper-competitive global market, the question of timing must be considered. Only time will tell whether or not the changes emerging in Japan's institutions will be fast enough or comprehensive enough to ensure a sustainable competitive advantage, and whether the institutions in the USA are capable of remaining resilient amid the turmoil of the 21st century. Similarly, the ongoing transformation of Chinese institutions and its global repercussions remains little understood in an increasingly global and digital marketplace.

References

Bisin A, Verdier T (2000) Beyond the melting pot: Cultural transmission, marriage, and the evolution of ethnic and religions traits. The Quarterly Journal of Economics 115(3):955–988

Blili S, Raymond L (1993) Information technology: Threats and opportunities for small- and medium-sized enterprises. International Journal of Information Management 13(6):439–448

Boisot M, Child J (1996) From fiefs to clans and network capitalism: Explaining China's emerging economic order. Administrative Science Quarterly 41:600–628

Bridge J, Peel M J (1999) A study of computer usage and strategic planning in SME sector. International Small Business Journal 17(4):82–87

CNNIC (2005) 15th statistical survey report on the Internet development in China, CNNIC, Beijing http://www.cnnic.net.cn

COMTEX (2004) Finance CustomWire, COMTEX, Alexandria

Covin J G, Slevin D P (1989) Strategic management of small firms in hostile and benign environments. Strategic Management Journal 10(1):75–87

Delone W H (1988) Determinants of success for computer usage in small business. MIS Quarterly 12(1):51–61

Dodgeson M (1993) Learning, trust and technological collaboration. Human Relations 46(1):77–95

Gao B (2001) Japan's economic dilemma: The institutional origin of prosperity and stagnation. Cambridge University Press, Cambridge

Griffy-Brown C, Watanabe C (1999) Inter-firm technology spillover and the creation of a 'virtuous cycle' between R&D, market growth and price reduction: The case of photovoltaic power generation (PV) development in Japan. International Greenbook Proceedings from the International Workshop on Technological Change and the Environment, International Institute of Applied Systems Analysis, Laxenburg, 21–22 June, 1999

Griffy-Brown C, Nagamatsu A, Watanabe C, Zhu B (2002) Technology spillovers and economic vitality: An analysis of institutional flexibility in Japan with comparison to the US. International Journal of Technology Management 23(7–8):746–768

Grossman G M, Helpman E (1991) Innovation and growth in the global economy. MIT Press, Cambridge

Henderson J C, Venkatraman N (1999) Strategic alignment: Leveraging information technology for transforming organizations. IBM Systems Journal 38:472–484

Hollingsworth J R, Boyer R (1997) Contemporary capitalism: The embeddedness of institutions. Cambridge University Press, Cambridge

JSBRI (2004) White paper on small and medium enterprises in Japan 2004—The limitless potential of the diversity of small and medium enterprises. JSBRI, Tokyo
Katz R (1998) Japan: The system that soured. ME Sharpe, New York
Kyobe M (2004) Investigating the strategic utilisation of IT resources in the small and medium-sized firms of the Eastern Free State province. International Small Business Journal 22(2):131–158
Li S, Vertinsky I, Zhou D (2004) The emergence of private ownership in China. Journal of Business Research 57(19):1145–1152
MacRae H (1995) The world in 2020: Power, culture and prosperity. Harvard Business School Press, Boston
Mintzberg H (1990) Strategy formation: Schools of thought. In: Fredrickson J W (ed) Perspectives on strategic management, HarperBusiness, New York
National Bureau of Statistics (1992a–1999a) China statistical yearbook. NBS, Beijing
National Bureau of Statistics (1993b–1996b) Database of industrial firms. NBS, Beijing
National Bureau of Statistics (1994a) State statistical survey and report systems. NBS, Beijing
Neidleman L D (1979) Computer usage by small and medium sized European firms: An empirical study. Information & Management 2(2):67–77
Ng C Y, Griffy-Brown C (1999) (eds) Trends and issues in East Asia. Foundation for Advanced Studies in International Development, Japanese Ministry of Foreign Affairs, Tokyo
Ng C Y, Griffy-Brown C (2001) (eds) Knowledge-based economic growth and socio-political implications in East Asia. Foundation for Advanced Studies in International Development, Tokyo
North D C (1994) Economic performance through time. The American Economic Review 84(3):359–368
OECD (1997) Special issue on information infrastructures. STI Review, OECD, Paris
OECD (2001) The new economy: Beyond the hype. Final report on the OECD Growth Project, OECD, Paris
Pempel T J (1998) Regime shift: Comparative dynamics of the Japanese political economy. Cornell University Press, Ithaca
Peterson R (ed) (1997) Electronic marketing and the consumer. Sage Publications, London
Qian Y, Weingast B R (1997) Federalism as a commitment to preserving market incentives. Journal of Economic Perspectives 11(4):83–92
Romer P M (1986) Increasing returns and long-run growth. Journal of Political Economy 94(5):1002–1037
Romer P M (1994) Beyond classical and Keynesian macroeconomic policy. Policy Options 15(6):15–21
Sabherwal R, King W R (1991) An empirical taxonomy of the decision-making processes concerning strategic applications of information systems. Journal of Information Systems 11(4):177–214

Senge P M (1990) The fifth discipline: The art & practice of the learning organization. Doubleday, New York

Stata R (1989) Organizational learning—The key to management innovation. Sloan Management Review 30(3):63–74

Tan Z, Clark T (2000) Internet diffusion in the USA and China. INFO 2(6):595–604

Tan Z, Ouyang W (2002) Globalization and e-commerce I: Factors affecting e-commerce diffusion in China. Communications of the Association for Information Systems 10:4–32

Telecommunications Council (2000) The info-communications vision for the 21st century. Telecommunications Council for the Minister of Posts and Telecommunications, Tokyo

US Bureau of Census (2004) E-commerce multi-sector report, US Bureau of Census, Washington, D.C.

Walder A G (1995) Local governments as industrial firms: An organizational analysis of China's transitional economy. American Journal of Sociology 101(2):263–301

Watanabe C (2002) Industrial ecology and technology policy: Japanese experience. In: Ayres R U, Ayres L M (eds) A handbook of industrial ecology, Edward Elgar, Cheltenham

Watanabe C, Kondo R, Ouchi N, Wei H (2003) Formation of IT features through interaction with institutional systems: Empirical evidence of unique epidemic behavior. Technovation 23(3):205–219

Yin R K (1994) Case study research: Design and methods. 2nd edn. Sage Publishing, Beverly Hills

Analyzing the Export Intensity of Electronics Firms in India*

Vinnie Jauhari

Abstract The second half of the 1990s witnessed an almost threefold increase in the number of exports by India's electronics industry. This chapter proposes and empirically tests a model to analyze the export intensity of 164 electronics firms in India; the Tobit model is estimated, using firm-level panel data from India's electronics industry for the period 2000–2005. The results indicate that firm size, foreign direct investment, and capital employed have played an important role in boosting exports in this sector. The findings have policy implications for the measures to be implemented to improve the electronics sector's performance in India and in other, similar countries.

Key words: Export Intensity, Electronics Industry, India, Institutional Context, Foreign Direct Investment, Empirical Analysis

1 Introduction

Different nations follow different growth trajectories. This may be attributed to a country's socio-cultural context, the institutional dynamism of various institutions operating within the given country, its level of economic development and economic regime, and the nature of its different institutions. Developed countries have relied on investments in education and R&D, especially in the high-tech area, to trigger the

* This is a slightly revised version of an article that was published in the International Journal of Innovation Management 11(3):379–396

Vinnie Jauhari
Region Lead, HP Labs Open Innovation Office, Tower D, 8th Floor, Mehrauli-Gurgaon Road, Gurgaon 122002, India, e-mail: vinnie.jauhari@hp.com

M. V. Van Geenhuizen et al., (eds). *Technological Innovation Across Nations*,
DOI: 10.1007/978-3-642-00158-1_5,

growth of firms. Asian countries, on the other hand, have implemented an export-oriented growth strategy. Different growth model options for emerging economies ought to be explored. For some countries, a higher degree of exports may result in a higher degree of competitiveness (Baldwin and Gu, 2004). The overall economic framework, the existing regulatory framework, foreign direct investment (FDI), and the nature of competition in a given country are all factors that contribute to a country's export orientation. Openness to trade may not be a panacea. Several studies document developing countries' differing experiences following the opening of their markets to trade. These divergences in experience derive from the differences in the composition of the countries, their approach to governance, the quality of their institutions, levels of education and health, and law and order (Blanke et al., 2003). Trade can be a catalyst for growth, but only when the appropriate institutions are in place. The general business environment is an important factor as well in determining the extent of benefits a country may reap.

India's closed market economy and licensing regime impeded the development of excellence in manufacturing for many years. Export performance can be considered one of the indices of productivity. In Porter's framework, countries move through three stages of development during which they produce increasingly sophisticated products, namely the factor-driven stage, the investment-driven stage, and the innovation-driven stage (Porter, 1990). In each phase, national competitive advantage is conditional on different determinants, which may also drive national trade performance. The factor-driven stage rests on a country's relative resources, such as abundance of low-cost labor. In the investment-driven stage, competitive advantage is determined by higher levels of efficiency in the production of standard goods, with an increase in investments during the growth in productivity. And finally, in the innovation-driven stage, competitive advantage derives from a country's innovative capacity, particularly in the production of high-technology products. Several segments of India's electronics sector have reached the investment-driven stage.

2 Objectives

The objective of this chapter is to assess the role of the factors that influence the export intensity of India's electronics industry. A model of export intensity is proposed and tested empirically. The objective is to assess the impact of certain firm-level variables on the export orientation of firms in the electronics industry.

Such a study is particularly useful, considering that the electronics sector grew threefold in the late 1990s and continues to grow exponentially, especially India's software sector. At the global level, the IT, consumer electronics, and telecommunications sector is one of the most important in terms of value-added. International trade in this sector was estimated at around \$632 billion in 2000, or approximately 11% of total world trade (The Global Competitiveness Report 2002–2003).

3 Review of the Literature

An industry's export orientation may drive growth in an economy. The literature suggests that several variables have an impact on a firm's export intensity. Banga (2003) asserts that a larger firm size, greater capital intensity, and more exposure to foreign markets contribute to higher export intensity. Banga's findings also indicate that enterprises with advanced technology resources and a superior knowledge base are more likely to be export-oriented than other enterprises. An abundance of research exists suggesting that technology is an important factor in explaining export performance and competitiveness (Gruber et al., 1967; Caves et al., 1980; Soete, 1981, 1987). Hughes (1986) found that the export intensity of industries in the UK was linked in particular to the extent of the industries' R&D investment and available skills. Hasan and Raturi (2003) used firm-level data from India's manufacturing industry to explore the determinants of export intensity, focusing on the role played by technology. They concluded that investments in technology through R&D and technology transfer agreements have facilitated the entry of Indian firms into the export market. However, their influence on the volume of exports is fairly limited. Factors that have a far-reaching influence on both export participation and volume include labor intensity and, in particular, firm size.

Baldwin and Gu (2004) explored the impact of trade liberalization on exports. They also examined the link between exports and innovation at the firm level. In their study on Canadian enterprises, they concluded that trade liberalization was a key factor behind the strong export growth of Canada's manufacturing sector. Furthermore, export market participation was associated with higher labor productivity growth. Their study provides evidence suggesting that export market participation is linked with industry specialization, and that exports are connected to higher levels of knowledge and skills and exposure to international competition. The findings also indicated that exporting facilitates the transfer of knowledge across countries and had enhanced the Canadian economy's innovation process. Finally, increased R&D collaboration with foreign buyers exposed Canadian enterprises to foreign technologies.

Czinkota (1996) contends that firms benefit from market diversification through the mechanism of exporting. Exporting also enables firms to learn from competition, sensitizes them to different demand structures and cultural dimensions, and demonstrates a firm's ability to persevere in less familiar environments. As a result, the firm becomes a stronger competitor at home. For their study, Lee and Blevins (1990) identified several variables as determinants of performance across different countries. These variables included firm size, diversification, advertising intensity, export activity, capital intensity, debt ratios, inventory, labor productivity, investment activity, and plant equipment. The impact of FDI on host economies can be analyzed in the context of its effects on growth-driving factors such as investment, human capital, exports, and technology. Recent studies suggest that FDI has the potential to enhance the economic growth of host countries through spillover effects and technology transfer. Spillover effects occur when advanced technologies and managerial skills in the form of FDI are diffused to domestic companies, solely on account of the presence of multinational firms (Zhang, 2001). The technology and productivity of local firms may improve as FDI creates backward and forward linkages, and foreign firms provide technical assistance to their local suppliers and customers (Rodriguez-Clare, 1996). Ozawa (1992) introduced FDI into Porter's theory, which both affects and reflects the position of a national economy. Ozawa proposed the following relationship, as indicated in Fig. 1.

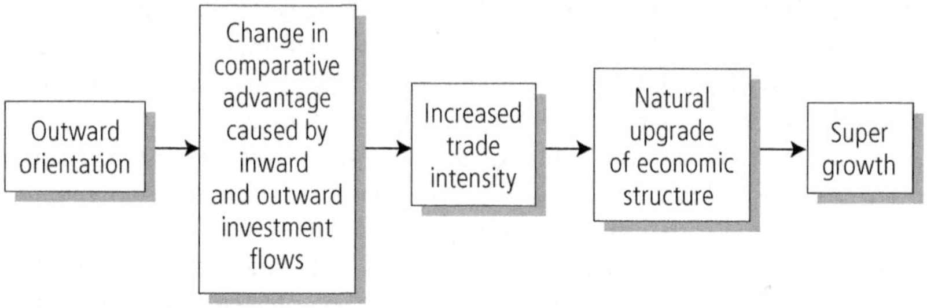

Fig. 1. Increased Foreign Trade as a Result of FDI
Source: Wysokinska (1998)

Kumar and Siddharthan (1993) analyzed the links between export behavior, technology, and firm size in industrialized and developing countries. A wide disparity in export performance across firms within a particular industry suggests that firm characteristics also play an important role. It has been argued that the technology factor may explain the export performance of India's small- and medium-sized technology enterprises.

FDI and increased trade in high-tech products offer developing countries the possibility of benefiting from the positive effects of technology diffusion. According to UNCTAD's World Investment Report 2002, Transnational Corporations (TNCs) can play a key role in improving developing countries' export performance through FDI. TNCs can complement a country's own resources and capabilities, and provide some of the lacking elements for greater competitiveness (UNCTAD, 2002). FDI and increased trade in high-tech products generate the positive effects of technology diffusion. However, to what extent certain countries actually benefit from technological diffusion is the subject of intense debate. If these countries were only selected because of their low-cost labor and for the purpose of re-exporting, technology diffusion to the rest of the economy would be more problematic (Blanke et al., 2003).

3.1 *Proposed Model for Exploring Export Intensity of Electronics Firms in India*

The proposed model for evaluating export intensity is described below:

Export intensity = F (the firm's size, its R&D intensity, advertisement intensity, capital imports, imports of spares and stores, payment of royalty and technical fees, capital labor ratio, capital output ratio, and FDI). This can be expressed by the following mathematical equation:

$$Y = a + b1X1 + b2X2 + b3X3 + b4X4 + b5X5 + b6X6 + b7X7 + b8X8 + b9X9$$

3.1.1 Dependent Variable

Y = Export Intensity (exports/sales)

3.1.2 Independent Variables

$X1$ Size of the firm (measured by log sales)
$X2$ R&D intensity (R&D expenditure/sales)
$X3$ Advertisement intensity of the firm (advertisement expenses/sales)
$X4$ Capital imports of the firm (embodied technology imports)
$X5$ Import of spares and stores (spares and stores are equipment and parts used in manufacturing) (embodied technology)
$X6$ Payment of royalty
$X7$ Capital labor ratio (K/L) (the proportion is used as a measure of a firm's capital accumulation)
$X8$ Capital output ratio (K/O)
$X9$ FDI (the firm's foreign equity)

These variables have been used in previous studies by Banga (2003).

3.2 Proposed Hypotheses for the Study

This section proposes a model to study the effect of different variables on a firm's export intensity. Factors affecting the export intensity of firms within an industry are examined within a simultaneous framework. The following hypotheses are tested:

1. The size of a firm has a positive influence on its export intensity.
2. A firm's R&D intensity has a positive influence on its export intensity.
3. The advertisement intensity of a firm has a positive influence on its export intensity.
4. A firm's capital imports have a positive influence on its export intensity.
5. The import of technology components has a positive influence on the firm's export intensity.
6. The payment of royalty has a positive influence on the export intensity of the firm.

7. The firm's capital labor ratio has a positive influence on its export intensity.
8. The capital output ratio of a firm has a positive influence on its export intensity.
9. FDI has a positive influence on a firm's export intensity.

All variables seem to have a plausible impact on firms' export intensity, reflecting the results of earlier studies in various countries. The variables that influence a firm's export intensity were derived from a review of the literature in this field.

4 Methodology

To determine export intensity, firm-level panel data was collected from the Capitaline Plus Database published in India and used for the firm-level analysis. The firm-level analysis here is based on the electronics firms' panel data for the period 2000–2005. The sample comprises a total of 164 firms, of which 116 are domestic firms and 48 have a foreign equity of more than 10% (henceforth referred to as 'foreign firms'). The firms included in the sample have a turnover exceeding $2.1 million. That is, only firms with a market presence, with the potential of entering the export market, were selected. The sample includes firms manufacturing electronics components, printed circuit boards (PCBs), control, instrumentation and broadcasting equipment, hardware and software, peripherals, and consumer electronics.

For the data analysis, the Tobit model was applied using limits (0, 1) by taking the averages for various variables for the period between 2000 and 2005. The Tobit model is particularly useful here, because many firms may have a value of zero for exports, a condition that is a basic assumption in Tobit analysis. These companies may presently not be exporting their products, but could begin exporting them in the near future. A test of multicollinearity was also performed before the actual analysis was carried out.

5 The Institutional Context of India's Electronics Industry

Globally, industrialized countries currently account for approximately 62% of world trade in the IT and consumer electronics sector. Japan demonstrates the strongest export performance (Global Competitiveness Report 2002–2003). The USA, with a slightly higher market share than Japan, ranks lower in the Trade Performance Index (The Global Competitiveness Report 2002–2003)[2], because of the USA's weak

[2] The Trade Performance Index (source: The Global Competitiveness Report 2002–2003). Trade competitiveness of specific countries can be measured in two ways, either through the trade performance current index (which measures a country's current level of competitiveness in a given sector) or the trade performance change index (which measures how a country's competitive position has changed over a period of five years). For further details, please refer to the Global Competitiveness Report 2002–2003.

performance in per capita terms and its negative net exports (the USA imports significantly more IT and consumer electronics products than it exports).

In the early 1970s, the electronics industry in India was predominantly labor intensive. The electronics industry comprises consumer electronics, instrumentation and industrial electronics, computers, communication and broadcast equipment, strategic electronics, and components and software. Emphasis was placed on using local technology and electronics. In the 1990s, the focus shifted toward maximizing indigenous production of electronics equipment for defense, communication, and space and atomic energy needs. Production units were spread across the country. The eighth five-year plan of 1991 outlined the objective to achieve international costs and prices. The electronics industry witnessed a cumulative annual growth rate of 20% and an increase in exports of over 40%. The years 1998–2002 were characterized by export, employment, and productivity-linked growth. The electronics and IT industry's export performance rose from approximately $1 million in 1970 to around $11.4 billion in 2002–2003 (Department of Information Technology 2004). The exports in this segment had not only increased in monetary terms, but also in terms of product range. The software industry, which was valued at around $139 million in 1991–1992 and accounted for only 9% of total electronics exports, recorded a production turnover of $8.1 billion in 2000–2001 and has surpassed electronics production.

The institutional system in India has played a distinct role in fostering the achievement of a high export level. The government has implemented a progressive policy since 1991, when liberalization measures were introduced. Prior to 1991, the licensing regime, which was characterized by restrictive trade practices, had resulted in several bottlenecks. The opening up of India's economy in 1991 led to more liberal imports of technology and other equipment. Industrial licensing has virtually been abolished in the electronics sector, except for the manufacture of electronic aerospace and defense equipment. These developments attracted FDI, leading to the production of more competitive products and an emphasis on exporting. The government initiated numerous measures to ensure a competitive infrastructure and the availability of funding opportunities. Export processing zones have been set up in various states, especially in Southern India. Software Technology Parks and Electronics Hardware Technology Parks have also been established. Until 1970, the electronics industry was concentrated around three centers—Bangalore, Bombay/Pune, and Delhi. To encourage a geographical spread, India's states were encouraged to establish electronics development corporations. Special incentives, such as sales tax concessions, the creation of separate industrial areas for electronics, subsidies for setting up units in rural areas, and financial participation either as a joint or assisted sector, were offered.

The electronics industry in India has grown through the joint efforts of both the private and public sectors. Fig. 2 illustrates the division of the sector in terms of ownership.

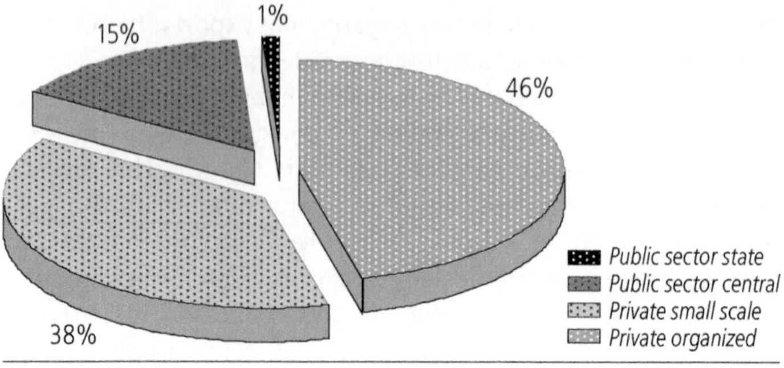

Fig. 2. India's Electronics Sector

6 Division of Electronics Output between the Public and Private Sector

Initially, the production of strategic electronics and communication in India was mainly reserved for the public sector. In the 1990s, the sector opened to private entrepreneurs as well. The private sector has played a particularly dynamic role in the software segment by setting up globally renowned firms such as Tata Consultancy Services, Infosys, and Wipro. These firms have been key drivers in exports from India's software segment. Fig. 3 delineates exports from various segments in India's electronics industry. As the figure illustrates, software exports significantly contribute to the total of exports from this sector. In the software segment, information technology enabled services have been the main driving force behind exports. India has been attracting a great deal of outsourcing investments, especially in the call center business, for both lower-end services such as customer support, and higher-end services like knowledge processing outsourcing for R&D and clinical trials. Competent manpower and cost advantages have been instrumental in generating exports from this sector. This, coupled with a progressive government policy and support, has been responsible for the growth of this segment.

India also has a very well-established education infrastructure with a strong educational base in technical training. The leading institutions of higher education, such as the Indian Institutes of Technology, are complemented by a number of regional engineering colleges and polytechnics. India furthermore has a strong R&D base in electronics through various laboratories associated with the Department of Space, Department of Information Technology, Department of Atomic Energy, Department of Defense R&D, and the Department of Scientific and Industrial R&D. Indian students receive prolific training in science and mathematics, and have good English language skills, which makes it easier to cater to global demands. Most importantly, India has a large, technically-trained manpower base.

6.1 Comparative Segment Exports for the Years 1993–2002

Fig. 3 illustrates the amount of exports from the electronics sector as a whole. The key driving force behind exports in this sector is the software segment, with the hardware and other segments lagging behind. The government policy and supporting infrastructure have played a tremendously important role in promoting exports from the software segment.

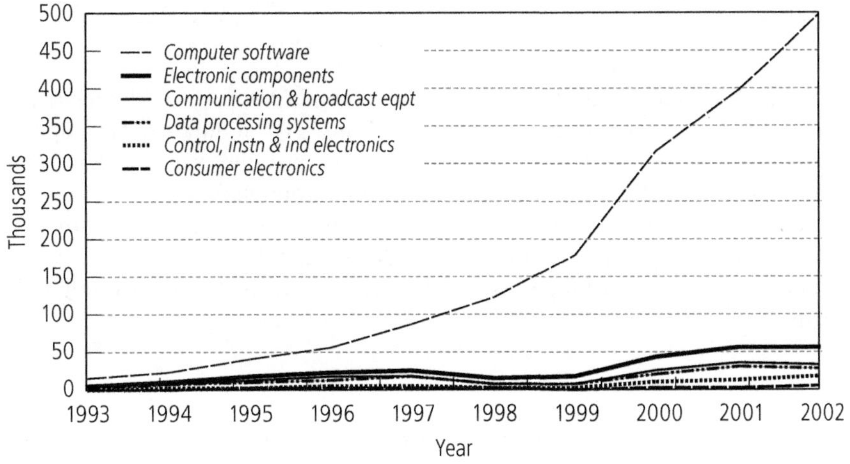

Fig. 3. Exports from India's Electronics Segments (INR 10 million ∼ $ 215.300)

6.2 Findings and Analysis

The overall analysis of the sector reveals some interesting insights. Table 1 provides a comparative look at firms with no foreign equity participation (henceforth referred to as 'domestic firms') and firms with 10% or more foreign equity participation (referred to as 'foreign firms').

The average sales by electrical and electronics firms in India is depicted in Table 1. Foreign firms contribute 67% of total production. The size of foreign firms is larger than that of domestic ones (which is statistically significant). Foreign firms' export intensity is also higher. The advertisement intensity of domestic companies is higher than that of foreign enterprises, perhaps because they are still in the process of building brands which is a resource-intensive exercise. The foreign brands have a higher recall value and brand equity due to their global presence. The R&D intensity of domestic firms is higher than that of foreign ones, while the import of spares by foreign firms is higher. They also outsource more in the initial stages before beginning to indigenize their operations.

Table 1. Summary of Averages for Various Variables

	Sales	Size (log sales)	Export intensity	Advertising intensity	R&D intensity	FDI	Imports of spares	Imports of capital equipment	Imports of royalty payments	Capital labour ratio	Capital output ratio
Foreign	168.64	3.49	0.13	0.01	0.05	551.52	3.72	4.31	0.76	30.70	7.43
Domestic	78.87	2.23	0.05	0.03	0.00	0	0.77	1.07	0.26	71.74	14.18

Foreign firms have higher imports of capital goods compared to domestic firms; in order to access global technology, foreign firms rely on imports. The capital labor ratio for foreign companies is also higher, because the wages and salaries of employees in domestic firms may be higher than in foreign firms.

The capital output ratio for foreign firms is higher, which implies that they have a lower output than domestic firms for the same capital employed. This indicates that domestic firms utilize their resources more effectively than their foreign counterparts.

6.3 Tobit Results for Electronics Firms in India

The results of the Tobit analysis for the 164 firms included in the study are indicated in Table 2. The firms' export intensity represents the dependent variable.

The results reveal that the following variables affect the electronics sector's export intensity:

⇒ Size of the firm (measured by its log sales);
⇒ FDI;
⇒ Capital output ratio.

The size of a given firm has an impact on its export intensity. The larger a firm, the larger the resources available for employing better technology and investing in quality standards. FDI intensity ensures access to current technologies. Spillover of learning techniques is likely, and the expertise brought in by foreign investors may lead to better management practices. Moreover, training and support may be offered and the foreign firm's systems and procedures adopted by the domestic firm. This result contrasts with those from some earlier studies on the impact of FDI on exports. The empirical results of Aggarwal and Goldar (1999) showed that in a simultaneous framework determining the intensity of technology growth, growth of a firm, level of technical efficiency, and export intensity, the foreign equity participation variable was not statistically significant in determining a firm's export intensity. Studies on India have found that export-oriented industries have not attracted FDI and that FDI has in fact little impact on exports from India (Siddharthan, 1999; Aggarwal, 2000; Kumar, 1998; Kumar and Siddharthan, 1997). FDI in India's manufacturing sector has been domestic market-oriented and not efficiency-seeking in nature (Sharma, 2000; Kumar, 1995; Aggarwal and Goldar, 1999). However, Parameswaran's research points out that firms' involvement in international trade through the export and import of raw materials and technology has a positive effect on technical efficiency (Parameswaran, 2002).

Capital ratio is also a determinant of export intensity. A higher degree of capital employed has an influence on a firm's export intensity. Larger firms with more available capital are better prepared to export their products and thus better understand international markets.

Table 2. Results of the Variables Affecting the Export Intensity of 164 Firms

Variable	Coefficient	t value	Level of Significance
Log sales	0.171	3.19	0.002
FDI	0.005	2.63	0.009
Imports of spares	–0.0017	–2.08	0.04
Imports of capital goods	0.000	–0.07	0.943
Royalty	0.003	0.35	0.724
K/L	–0.000	–0.98	0.330
K/O	0.144	6.51	0.000
Advertising intensity	–3.668	–1.45	0.148
R&D intensity	0.004	0.12	0.905
Constant	–2.13	–1.98	0.050
SE	0.34		

Limited dependent model—Censored
Number of observations—164
Log likelihood function—71.91

6.4 *Export Intensity of Domestic Firms*

The results for Indian firms are presented in Table 3.

Table 3 indicates that the following variables have an impact on export intensity:

⇒ Size of the firm;
⇒ Capital/labor ratio;
⇒ Advertising intensity.

Among domestic firms, size is a determinant of export intensity. A larger size has an impact on the firm's ability to procure appropriate technology, recruit the right kind of employees, and invest in marketing and marketing-related activities. The figures for capital labor ratio signify that higher capital infusion is required. Domestic firms better utilize labor, resulting in higher productivity compared to foreign firms, as indicated in Table 1 as well which compares foreign and domestic firms.

Table 3. Results for the Tobit Analysis of Domestic Firms

Variable	Coefficient	t value	Level of Significance
Log sales	0.186	2.56	0.012
FDI	0.024	1.76	0.081
Imports of spares	–0.001	–1.22	0.225
Imports of capital goods	0.009	1.85	0.067
Royalty	0.000	0.04	0.967
K/L	–0.003	–2.49	0.014
K/O	0.197	6.73	0.000
Advertising intensity	–4.609	–1.68	0.096
R&D intensity	0.000	0.01	0.996
Constant	–0.308	–2.26	0.026
SE	0.327		

Tobit model limits 0, 1
Limited dependent variable model—Censored
Number of observations—116
Log likelihood function—48.57
Component score coefficient component

Advertising intensity has an impact on the export intensity of Indian firms. This has significant policy implications. The government needs to invest in enhancing the visibility of Indian brands. Globally, products manufactured in China have an edge over those manufactured in India. The Indian government and corporate firms need to think about ways to increase the visibility of their product range.

The technology index was prepared for domestic firms using principal component analysis. The purpose was to determine the cumulative effect of technology as a component consisting of variables such as R&D intensity, imports of stores and spares, imports of capital equipment, and royalty payments. The principal component analysis was carried out to examine the combined effect of all these variables on the export intensity of firms. The results are presented in Table 4.

Table 4. Component Score Coefficient Matrix*

Component	Score Coefficient Component
Imports of spares	1.00
Imports of capital goods	0.10
Royalty	0.43
R&D intensity	0.83

*Used principal component analysis, rotation method: quartimax with Kaiser normalization

When the Tobit censored model was used again for Indian firms from different electronics segments using the aggregate technology variable based on the above results, it was found that the size and technology component were important determinants of these firms' export intensity. These results are presented in Table 5.

Table 5. Impact of Technology and Size on Export Technology

Variable	Coefficient	t value	Level of significance
Log sales	0.164	3.11	0.00
FDI	0.006	2.75	0.00
Imports of spares	0.00	0.00	0.90
Imports of capital goods	0.00	0.12	0.90
K/L	–0.00	–0.97	0.33
K/O	0.148	6.65	0.00
Advertising intensity	3.14	1.25	0.213
R&D intensity	0.001	2.29	0.02
Constant	–0.22	–2.10	0.03
SE	0.34		

Number of observations 168
Log likelihood function—75.48

These results are in line with earlier research findings. There is abundant research suggesting that technology is an important factor in explaining export performance and competitiveness (Gruber et al., 1967; Caves et al., 1980; Soete, 1981, 1987). Hughes (1986) found the export intensity of industries in the UK to be closely linked with their R&D and skill intensities and, inversely, with the average R&D intensity of industries in the USA, France, Germany, and Japan. Few studies have examined the role of technology with regard to trade in developing countries. Lall (1986) concludes that R&D expenditure, and not its intensity, are significant, with a negative sign in the case of engineering firms and a positive sign for chemical firms. Royalty payments as a proxy for the extent of technology imports and skill intensity proved to be an insignificant coefficient throughout the analysis.

7 Conclusions and Implications

The results of the various analytical tools applied in the analysis of electronics firms in India indicate that not all of the proposed hypotheses are valid. It can be concluded that firm size, FDI, and capital investment have an impact on the export intensity of firms operating in the electronics sector. The technology component and the advertising intensity of domestic firms play an important role in India's electronics sector. This study has important implications for policy makers and entrepreneurs. Exports can foster innovation within an industry and thus promote its growth. The findings also point out that firm size, FDI, and capital investment have been the main driving forces behind exports from electronics firms in India. Investment in technology and other financial investments may increase a firm's competitiveness. Though SMEs are generally viewed as drivers of innovation, entrepreneurship at a small scale may not be sustainable in the case of the electronics sector. Substantial financial investments are required for state-of-the-art technology and marketing. Hence, firm size and investments in technology emerge as the driving force behind exports from Indian electronics firms. The government needs to consider provisions for longer-term loans to initiate sustainable entrepreneurship. Sustainable entrepreneurship in the electronics sector requires sizeable investments, which private entrepreneurs may not be willing to make. Here, venture capitalists could step in, as they are more open-minded with regard to funding and have higher risk-taking capabilities. The trend that is already visible in the software segment is that most venture capital is flowing into firms that have existed for several years and need more large-scale investments.

Technology spillover from foreign firms is also evident. Technology imports and foreign equity participation have led to a spread of know-how and have contributed to an enhanced export orientation of India's electronics firms. The divergent performance of the technology and capital intensity variables in high and lower technology industries is of considerable significance for policy. In low and medium technology industries, export competitiveness is achieved through indigenous technological know-how and the implementation of labor-intensive production processes. In the high technology industries, on the other hand, technology imports—that is, affiliation with multinational enterprises—as well as a higher degree of automation and modernization appear to be important prerequisites for entering international markets. As the technology variable appears to be a significant driver of export intensity, measures need to be taken to reduce the tariff barriers and ensure that the technology being used by domestic firms is up-to-date. The Chinese government, for example, is committed to supporting local entrepreneurship. The Indian government should consider such a policy as well.

Exports from local Indian firms must be encouraged. A 'single window system' could be set up which offers technology solutions for firms operating within a given industry. This would minimize the obstacles for entrepreneurs interested in making investments. Barriers in the setting up of enterprises deter entrepreneurship. Furthermore, though numerous agencies that provide required technology exist, certifications and approval from other agencies continue to be compulsory. All these formalities should be kept to a minimum. An entrepreneur has to frequently turn to multiple agencies for assistance, which complicates the decision-making procedure. These processes should become more user-friendly and uniform policy guidelines ought to be established. A regulatory mechanism for stipulating the timeframe by when all clearances must have either been approved or rejected must be implemented. Power, water, and telephone connections should be available on demand. Import, export, excise and sales tax formalities should be simplified so they can be completed in-house, and subject to random audits by regulating agencies.

The establishment of Technology Parks in India, especially the Software Export Promotion Parks, has been a step in this direction. India's software industry in particular has demonstrated that institutional support by way of providing a 'one roof solution' has worked quite well. A significant share of exports originates from the Software Parks in India.

If a policy of investments in infrastructure is implemented and procedures simplified, and if the government further facilitates access to state-of-the-art technology to support entrepreneurs' initiatives, the industry has the potential to leapfrog to the next stage of growth. The government's policy for the software industry is a case in point. The Indian government needs to make focused investments, so that resources for R&D, technology development, and export efforts are not thinly spread. The focus should be on new areas of technology, including New Wave Internet access devices and technologies, PCs for the masses, cable-, wireless- and broadband technologies, digital compression technologies, and G wireless technologies. Consortia consisting of government, venture capitalists, and financial institutions for R&D investment need to be encouraged.

The hardware and software industries will both have to grow. To reach the target of $50 billion in exports in 2008, IT penetration will have to be intensified. FDI needs to be encouraged in the hardware sector, and tariffs on input and capital goods need to be reduced considerably. Furthermore, the EXIM policy (the Indian government's policy on export and import) needs to be simplified and clear customs procedures introduced.

Indian firms will also have to undertake efforts to increase the visibility of their products. However, efforts at the national and industry levels are necessary in order to create a high degree of acceptance for electronics products 'made in India'. Such initiatives are already under way for the software sector, but the hardware and other equipment sectors have thus far not been a priority for the Indian government. The Indian government needs to therefore also support the indigenous production of hardware and hardware-related equipment. A competitive development of India's hardware sector is essential to sustain the country's growth in the future.

Future research could focus on collecting empirical data through interviews to provide greater insight into the issues currently preoccupying electronics firms and the industry.

References

Aggarwal A (2000) Liberalisation, MNE affiliates and export performance: Evidence from Indian manufacturing. Working Paper Series 19/2000, Institute for Economic Growth, New Delhi

Aggarwal R N, Goldar B N (1999) Technology imports, growth, efficiency and export performance of Indian engineering firms in the pre- and post-reform period. Working Paper Series E/201/99, Institute of Economic Growth, New Delhi

Baldwin J R, Gu W (2004) Trade liberalization: Export-market participation, productivity growth and innovation. Oxford Review of Economic Policy 20(3):372–392

Banga R (2003) The differential impact of Japanese and US foreign direct investments on exports of Indian manufacturing. Working Paper 106, Indian Council for Research on International Economic Relations, New Delhi

Blanke J, Kirshbach F V, Mimouni M, Pasteels J M (2003) Export performance and stages of development. The Global Competitiveness Report 2002–2003, Oxford University Press, New York

Caves R E, Porter M E, Spence A M, Scott J T (1980) Competition in the open economy: A model applied to Canada. Harvard University Press, Cambridge

Czinkota M R (1996) Why national export promotion? International Trade Forum, Geneva

Gruber W H, Mehta D, Vernon R (1967) The R&D factor in international trade and investment of US industries. Journal of Political Economy 75:20–37

Hasan R, Raturi M (2003) Does investing in technology affect exports? Evidence from Indian firms. Blackwell, Oxford

Hughes K (1986) Exports and technology. Cambridge University Press, Cambridge

Kumar N (1990) Multinational enterprises in India: Industrial distribution, characteristics and performance. Routledge, London

Kumar N, Siddharthan N S (1993) Technology, firm size and export behaviour in developing countries: The case of Indian enterprises. UNU/Intech Working Paper 9, Maastricht

Kumar N (1995) Industrialization, liberalization and two way flows of foreign direct investment: The case of India. Discussion Paper Series 9504

Kumar N, Siddharthan N S (1997) Technology market structure and internationalization. Routledge, New York

Kumar N (1998) Multinational enterprises, regional economic integration and export platform production in the host countries: An empirical analysis for US and Japanese corporations. Weltwirtschaftliches Archiv 134(3):450–483

Lall S (1986) Technological development and export performance in LDCs: Leading engineering and chemical firms in India. Weltwirtschaftliches Archiv 122:80–91

Lee J, Blevins D E (1990) Profitability and sales growth in industrialized versus newly industrializing countries. Management International Review 30(1):87–100

Ozawa T (1992) Foreign direct investment and economic development. Transnational Corporations 1(1):27–54

Parameswaran M (2002) Economic reforms and technical efficiency: Firm level evidence from selected industries in India. Working Paper 339, www.cds.edu

Porter M (1990) The competitive advantage of nations. Free Press, New York

Rodriguez-Clare A (1996) Multinationals, linkages and economic development. American Economic Review 86:852–873

Sharma K (2000) Export growth in India: Has FDI played a role? Discussion Paper 816, Economic Growth Center Yale University, New Haven

Siddharthan N S (1999) Technology transfer, WTO and emerging issues. Paper presented at the National Seminar on "Economy, Society and Polity in South Asia" Retrospect and Prospects at the Dawn of the Next Millennium, November 16–17, 1999 at The Institute of Economic Growth, New Delhi

Soete L (1981) A general test of technological gap theory. Weltwirtschaftliches Archiv 117:638–659

Soete L (1987) The impact of technological innovation on international trade patterns: The evidence reconsidered. Research Policy 16:101–130

UNCTAD (2002) World investment report 2002. Transnational Corporations and Export Competitiveness, UNCTAD, Geneva

Wysokinska Z (1998) Impact of foreign direct investment on export competitiveness. Russian & East European Finance and Trade 34(4):64–87

Zhang K H (2001) Does foreign direct investment promote economic growth? Evidence from East Asia and Latin America. Contemporary Economic Policy 19(2):175–185

Part II
Applied Studies: Institutional Systems, Entrepreneurship, Knowledge Transfer and Learning

Growth of Technology Incubators: An Evolutionary Perspective

Marina van Geenhuizen and Danny Soetanto

Abstract Since the introduction of the first university-based technology incubator, this policy tool has gradually transformed under the influence of different (local) circumstances, particularly changes in the relationships between universities, government, and industry. In this chapter, we examine evolutionary changes within the institutional characteristics of technology incubators, and attempt to identify, among other things, the influence these changes have on the growth patterns of incubators. The three institutional characteristics considered include the model of stakeholder involvement (from single to multiple), the model of support (from conventional to value-added), and the overall growth strategy of the incubator (from focused to broad). We use a unique cross-national database of 40 incubators derived from a meta-analysis and apply a relatively new approach that corresponds with small, select samples and qualitative, sometimes fuzzy, data, namely, a rough set analysis. Overall, it appears that the combination of multiple stakeholder involvement and a location in a non-metropolitan area provides the most convincing explanation for strong growth. Conventional support coupled with a focused incubator strategy ranks second in terms of explanatory power, but relates to weak rather than strong growth.

Key words: University-Linked Incubators, Evolutionary Growth, Rough Set Analysis, Academic Spin-Off Companies

Marina van Geenhuizen
Faculty of Technology, Policy and Management, Delft University of Technology, PO Box 5015, 2600 GA Delft, The Netherlands, e-mail: m.s.vangeenhuizen@tudelft.nl

Danny Soetanto
Faculty of Technology, Policy and Management, Delft University of Technology, PO Box 5015, 2600 GA Delft, The Netherlands

M. V. Van Geenhuizen et al., (eds). *Technological Innovation Across Nations*,
DOI: 10.1007/978-3-642-00158-1_6,

1 Introduction

No other field in economic policy appears to receive more attention today than the knowledge-based economy. Since the launch of the Lisbon Strategy by the European Commission in 2000, the knowledge-based economy has been at the top of virtually all national European policy agendas. Governments advocate knowledge as the major engine of economic growth and global competitiveness, and entrepreneurship as a conduit for knowledge spillovers and commercialization. Investments in new knowledge and ideas do not necessarily automatically spill over and stimulate commercialization (Audretsch et al., 2006). Many hurdles must be overcome before the fruits of new knowledge can be reaped, most of which originate from the institutional system that often lags behind (Freeman and Perez, 1988), as exemplified by restrictive regulation in the application of new biotechnology. It is the interplay between governments, entrepreneurs, technology, and related institutions that determines how quickly and to what extent new knowledge leads to economic growth, and vice versa (Aghion and Howitt, 1998). One specific reason for governments to be involved in the knowledge-based economy is to mitigate the impacts of various market flaws that hamper the growth of small high-technology companies, like information asymmetry, equity gap, or the lack of dedicated rental space. In this context, both national and regional governments have designed a whole range of policy instruments, including R&D subsidies, rental space for science parks and incubators, and intermediary organizations. Approaches and measures to establish or improve the knowledge economy, particularly at the regional level, have received increased attention with a number of policies implemented in many regions. However, systematic knowledge necessary to support relevant decisions regarding the adequacy of policy tools and practices remains scarce.

One way to support the creation of the knowledge-based economy is through public investment in universities and organizational structures that foster knowledge spillovers into the regional economy. Today's perceived role of universities goes far beyond the classical model of increasing the level of scientific knowledge, educating young adults, supplying the labor market with new graduates, and acting as a source of multiplier effects in the regional economy (Armstrong and Taylor, 2001). At present, strong emphasis is placed on the economic utilization of academic knowledge (Varga, 1998), the creation of new employment by academic spin-off companies, and on universities taking an entrepreneurial role (Boucher et al., 2003; Rothaermel et al., 2007; Shane, 2004). In this context, many universities, usually in collaboration with other stakeholders, have established incubators.

Although incubators have been recognized as being an effective tool for nurturing high-technology start-up companies, as with the learning region approaches (Morgan, 1997), little attention has been given to factors that systematically explain the different growth patterns of incubators from a quantitative and cross-national comparative perspective. Research undertaken to date has relied heavily on the qualitative analysis of a few case studies (Autio and Klofsten, 1998; Mian, 1997; Hannon and Chaplin, 2003), often with a focus on one particular country (Lockett et al., 2003). In addition, the most common stakeholders involved in such studies, namely, governments and universities, are non-profit organizations that have never felt a strong need to carry out a broad evaluation of the policy on technology incubators. Even if incubators were operated as public-private partnerships, evaluating them would be rather complicated due to lacking theoretical knowledge to characterize the precise nature of the business models and managerial practices involved (Phan et al., 2005). The virtual absence of quantitative and cross-national comparative studies is attributable to the fact that data are scarce and, if available, tend often to be fairly imprecise and biased. Consequently, this chapter aims to determine the different growth dynamics of incubators in various countries by using a meta-analytical approach and by applying a method that accommodates relatively imprecise (or 'fuzzy') qualitative data, i.e., a rough set analysis. Specifically, we address the following questions: to what extent do incubators follow different growth patterns, and how is that connected to the evolution of incubator models? Which other factors play a role in causing different growth patterns?

The chapter is structured as follows: first, we take a closer look at changes that have occurred both in the environment of incubators and within the incubators from a coevolutionary development perspective. We examine various incubator models in greater detail to arrive at a number of assumptions regarding their varying growth rates. Secondly, we describe the research design of the study which is based on a broad "causal" analysis, i.e., a rough set analysis. This is followed by an examination of the research findings. The chapter concludes with a discussion and evaluation of the major findings and proposes policy recommendations and topics for further research. We do not intend to investigate the success or failure of incubators. We also do not address the achievement of particular policy aims regarding the number of new spin-offs, nor do we answer the question whether there would have been any difference in the survival rate of spin-offs, if the latter had not been incubated. We will focus solely on the question why incubator development is more dynamic in some circumstances and less in others.

2 The Evolutionary Development of Incubators

An incubator is an apparatus that provides controlled environmental conditions to hatch eggs, cultivate microorganisms, or care for and protect premature or sick babies (Phan et al., 2005). Analogous to this, we define incubators primarily as organizational structures established by universities and other actors to facilitate and enhance the formation and growth of companies that deliver academic knowledge to the market. We do not perceive incubators strictly as real-estate or property-based initiatives providing businesses with a rental space. However, incubator organizations usually occupy one or more buildings that offer customized rooms and support (administrative) services. Our definition interprets incubators as institutions embedded within a range of other institutions and relationships, and as subject to change under their respective influence (Edquist and Johnson, 1997). Evolutionary processes of change are construed as step-by-step or gradual transformations.

University-industry-government relations can be regarded as adaptive networks that have gradually evolved over time. The networks fulfill three core functions, i.e., knowledge production, wealth generation, and governance (control). One of the most noticeable changes over time is a blurring of the boundaries of the functions of all three actors. Thus, in Western Europe and North America, universities have become increasingly entrepreneurial, while large companies have initiated knowledge production-based education on campuses by hosting training in specific academic fields. Governments for their part are increasingly engaged in enhancing the knowledge-based economy by improving conditions for entrepreneurial innovation (Etzkovitz, 2002). The role of universities has changed in the past decade; next to their traditional tasks in education and research, universities now engage in additional tasks in the entrepreneurial sphere. Universities have also adopted greater societal responsibility, such as addressing particular problems in the regional economy and society (Castells and Hall, 1994; Charles, 2003). There is no end in sight to such changes, implying that a continued integration and hybridization of functions awaits us in the coming years, even if the onset and pace of change varies in different countries (Viale and Etzkowitz, 2005). The nature of knowledge production itself has also changed over time in Western Europe and North America. The participation of knowledge producers in networks to support their activities has increased, and the emergence of new actors from outside universities and established research institutes plays a growing role in knowledge-producing teams, if often on a temporary, project basis only. In addition, there has been a shift to a mode of knowledge production in which research problems are situated across disciplinary boundaries, and where practical applicability of the research results is considered as important as scientific innovation (Etzkovitz, 2002; Nowotny et al., 2001). Finally, the emergence of university-industry-government relations in Asia introduced incubator models that involve greater emphasis on governance by the state.

The institutional changes mentioned above have manifested themselves differently throughout the developed world, leading to a large variation of incubator development with regard to size and growth dynamics (Aernoudt, 2004; Etzkovitz, 2002). However, we expect that other factors influence incubator growth as well. One such factor we identified in the literature is the regional (or urban) environment in terms of existing high-tech entrepreneurship and entrepreneurial culture (Clarysse et al., 2005; Roberts and Malone, 1996). Furthermore, the incubators' pace of organizational learning seems to play a role in the management and guidance of the incubation process and its output.

The first wave of incubators was established in the early 1980s and traditionally merged the concept of fostering new business development with that of technology transfer and commercialization (Phillips, 2002). First-generation incubators merely provided tangible support in the sense of physical and financial infrastructure, including affordable rental space, office services, laboratory equipment, and loans. Universities were often the only stakeholder in the first generation of incubators. Later, incubators began operating as entrepreneurial, non-profit organizations by bridging spin-offs with the required resources in light of broader stakeholder involvement (Hackett and Dilts, 2004). Additionally, support increasingly included *intangible* inputs, like training entrepreneurs how to access networks involving key actors, thereby improving their entrepreneurial skills (Maine and Garnsey, 2006). Social inputs are less well-defined and known than other inputs. Recently, new types of incubators with a focus on profit-seeking have emerged, indicating a broadening in the types of target companies with start-ups outside the university (Clarysse et al., 2005). Following the increased attention on social inputs, the emergence of the so-called "networked incubator" has been observed, in which support is focused on improving the skills of entrepreneurs to access strategically important network clusters to increase their legitimacy, skills, and resources (Bøllingtoft and Ulhøi, 2005; Hansen et al., 2000).

We summarize the abovementioned changes in the model of technology incubators as evolutionary shifts from a one-sided physical tool to an increasingly social and diversified, hybrid tool, moving across the boundaries of other areas, such as social capital. These shifts generally occur within three dimensions: 1) stakeholder involvement, 2) type of support, and 3) overall growth strategy of the incubator. The application of different incubator models across the developed world has resulted in a large variety of incubation practices and numerous new entrepreneurial ventures. Under which conditions incubators are most "productive", however, remains largely unknown.

3 Factors Influencing Incubator Growth

There are two related arguments in evolutionary thinking that guide our assumptions on the growth of incubators. First, according to evolutionary thinking, organizations need to adapt to changes in their environment quickly in order to survive and to remain competitive. Flexibility or adaptability of existing structures is of crucial importance (Bergh and Stagl, 2003). This implies that incubators which adapt new models of stakeholder involvement, overall strategy, and support in a timely fashion are more productive than others. Secondly, the development of new models has tended toward diversification, stemming from a growth trajectory that began with a "pure", one-goal initiative by a single stakeholder, and has gradually been influenced by a hybridization of the organizations and relationships involved. The diversification of stakeholders, of incubated start-ups, and of support may provide a more varied mix of resources to incubated start-ups which, to a certain extent, may contribute to a higher productivity of incubators.

Many authors have depicted stage models of new firm establishment and growth to understand the transformation from initial business start-ups to viable companies. In line with Clarysse et al. (2005), we broadly distinguish between three different stages, namely (1) the invention stage, faced with considerable technical uncertainty, (2) the transition stage, in which technical uncertainty decreases and the business idea is validated, and (3) the innovation stage in which growth expectations are validated. While moving through the various stages, spin-off companies face different resource needs that they can only partially meet on their own, prompting the need to rely on differentiated support from external resources over time (Reid and Garnsey, 1998). From the onset, these companies may already be highly diverse with divergent needs for resources, for example, with reference to the entrepreneur's previous experiences, the type and level of technology applied, and the type of product-markets perceived as opportunities (Druilhe and Garnsey, 2004; Maine and Garnsey, 2006; Mustar et al., 2006). Only recently has awareness of heterogeneity between academic spin-offs and the logical implication of this circumstance increased, namely the need of these spin-offs for a supply of differentiated support. In the next section, we will discuss various institutional factors related to the supply of support, next to other factors influencing the growth of incubators.

To a certain degree, incubator organizations attempt to imitate the creative setting of urban centers, where young, high-technology firms benefit from a large diversity of resources, including affordable rental space and knowledge spillovers (Leone and Struyck, 1976; Audretsch, 1998). Incubators are different with regard to the types of stakeholders involved, which leads to distinctions in creative atmosphere and wealth of resources. Some incubators are primarily organized by a university or research institute, while others internalize the triadic relationships between the university, government, and industry and the hybridization of their roles (Etzkovitz, 2002; Leydesdorff, 2003). The involvement of different stakeholders—as opposed to the involvement of just the university—enables potential access to more diverse networks and resources and the ability to better respond to the previously mentioned heterogeneity among incubated start-ups. Thus, we may assume that the model of multiple stakeholder involvement generates a more dynamic development compared to the model of single stakeholder involvement (Monck et al., 1988).

The next two institutional factors represent choices made by incubator organizations and their stakeholders, in other words, the overall incubation strategy and the model of support. Such choices are not always explicit and intentional. They often evolve on their own and are only consciously taken after several years of experience. Hannon and Chaplin (2003) identified two incubation strategies in the United Kingdom, pure incubators and flagship models. Pure incubators are traditionally established by universities and seek to exploit university potential by cultivating firms that commercialize research results. The flagship model, on the contrary, originates from initiatives by local or regional governments or real estate developers, and tends to be more profit-oriented. In this model, significant investment is required to initiate the project and revenue streams are necessary to support the running costs. Accordingly, the flagship strategy often urges incubators to attract an optimal number of new start-ups in order to operate in a financially sound way. More recently, new types of incubators have emerged that, like the flagship model, also pursue a certain level of profit and invite start-ups from outside the university to join the incubation program. These incubators are expected to grow faster than traditional ones because they have a larger capacity to absorb new entrants. More importantly, they benefit from input by a variety of knowledge sources and participate in more diversified networks to gather resources, matching the need for diverse support among the incubated spin-offs. Accordingly, incubators pursuing a certain degree of profit are likely to experience stronger growth than their "pure" counterparts. With regard to the model of support, we argue the following: conventional support primarily deals with the provision of tangible assets, such as spatial or laboratory facilities and financial support. However, an important evolution in support models is evident, namely from conventional to added-value, (i.e., social) support. The latter includes supporting start-ups by linking them to relevant networks, such as business angel networks, and introducing new methods in business mentoring and training to increase the spin-offs' capability to survive (Hannon and Chaplin, 2003; McAdam et al., 2006). Network support focuses on utilizing internal relationships—between the companies on the incubator site—and external relationships (see Cook et al., 2006, for details on cross-border networking in biotechnology incubators). Business mentoring and training may focus on building the capability of the entrepreneur to develop strategic thinking, to establish and maintain networks, and to negotiate with and win over potential customers. Active assistance in developing exit strategies may also be an element of added-value support, as in financial syndication and preparation of an initial public offering (IPO). We may assume that an emphasis on added-value support accelerates the growth of incubated spin-offs and reduces their nurturing time, thus increasing the number of incubated spin-offs per year.

In the previous discussion, we put forth the assumption that incubators which employ models reflecting diversity are more productive in terms of growth than other incubators. What falls beyond the scope of our analysis but is strongly related with strategy and support, are the target companies themselves. Clarysse et al. (2005) distinguish between three incubator models, namely the low selective model (attracting mainly self-employed start-ups), the supportive model (predominantly attracting start-ups that are economically attractive), and the incubator model (attracting start-ups of interest for financially profitable exit modes). These models have different requirements with regard to the resources needed by the incubated companies and to the support provided. However, since it is difficult to link this classification with the concept of diversity and to assess how these models work in reality, we do not include them in our analysis. In the remaining part of this section, we focus on three other incubator characteristics that potentially influence their pace of growth, including their experience and the environment in which they are located, their entrepreneurial spirit (i.e., attitude toward risk), and assets that can be utilized as resources (i.e., assets in the urban environment).

To manage an incubator, experience and professionalism in selection, monitoring, and coaching of start-ups appear to be critical (Smilor et al., 1988). A study in the USA reveals that it takes several years for incubators to mature in terms of gaining the capability to organize themselves efficiently, and produce independent and viable firms on a continuous basis (McKinnon and Hayhow, 1998). Experience-based learning, for example, improves an incubator management's capability to identify the most productive networks and to efficiently participate in them. Although learning is not a linear process, it increases with time. Accordingly, we may assume that incubators grow faster after gaining some initial experience. It must be noted that physical capacity indicators, like the incubator organization's building size or budget, remain beyond the scope of our analysis since we are interested in the analysis of *strategic* factors. For the same reason, operational details like access criteria and duration of incubator support are not included in our study.

With regard to the influence of entrepreneurial culture, we argue that uncertainty avoidance is an important dimension (Hofstede, 1991). Countries facing high levels of uncertainty avoidance generally reveal a tight regulation of entrepreneurship and a habitual resistance to embark on new, risky expeditions. By contrast, we assume that there is room for differentiation, experimentation, and risk-taking in business operations in countries with low levels of uncertainty avoidance. Such an entrepreneurial climate is also believed to stimulate incubator organizations to take more initiative and be more creative in their pursuit of new opportunities to support their ventures. All these factors may lead to a more dynamic growth of incubators.

The last aspect to be considered here involves differences in the quality of the incubators' urban environment, particularly the spatial concentration of human capital and knowledge. Jacobs (1961) draws attention to metropolitan cities which attract and mobilize talented and creative people. In a similar vein, a growing stream of more recent research has focused on studying factors that attract talented people (Glaeser et al., 2001; Florida, 2002). This closer examination has revealed that metropolitan cities which offer a high diversity of amenities, entertainment, and lifestyle, but also tolerance, safety, and social inclusion (in other words, soft infrastructure), have important advantages in attracting and retaining talented and creative people. In such an environment, a strong and experienced entrepreneurial community already exists, ready to take up the best projects and allocate appropriate resources. In addition, following the tradition of Marshallian thinking, large cities provide benefits from various agglomeration economies and supply-side externalities. The former include benefits ranging from easier access to, and reduced costs of, certain collective resources. The latter include positive externalities such as inputs from local supplier industries, labor-market pooling, and knowledge spillovers on the basis of tacit knowledge flows facilitated by proximity. Accordingly, Duranton and Puga (2003) contend that cooperation and sharing of indivisibilities, variety, specialization, and risk act as the leading benefits for innovative entrepreneurship in metropolitan areas. Given the previous ideas, we may assume that incubators in metropolitan cities enjoy the benefits of a large pool of talented entrepreneurs and of knowledge, as well as of the urban setting which acts as an incubator on its own. Accordingly, we expect such incubators to grow relatively fast in contrast to incubators in non-metropolitan settings, i.e., in rural areas and peripherally-located regions.

4 Structure of the Empirical Study

We used a unique database of incubators derived from a meta-analytical study, that is, a study based on an existing body of knowledge. A meta-analytical approach comprises the systematic application of a range of quantitative methods to assess common characteristics and variations across a set of separate but largely similar case studies on, more or less, the same phenomenon (Bergh et al., 1998; Cook et al., 1992). Meta-analysis makes synthesizing the results of many types of research possible, including opinion surveys, annual performance reports, experimental, and quasi-experimental studies. We included incubators' annual reports, articles in scientific and in popular journals, incubator manuals, as well as individual incubator Web sites. In the sampling procedure, we used various rules of thumb to specify the particular requirements the incubators had to fulfill: (1) *to be a technology incubator*; this type of incubator primarily supports technology-based start-ups and employs institutional links with and/or is located close to a university or research center; (2) *to cover a similar time period*; incubation processes in the 1980s cannot be compared with those of the 1990s because of the potential discrepancy in the influence of various macro-economic factors, a reason why we limited our study to one particular period, namely the years 1998 to 2002, and to countries with similar levels of macro-economic growth,[1] and (3) *to display particular characteristics of one (or more) of the factors discussed in the previous section*; the purpose was to have a substantial degree of variance in the conditions relating to these factors.

It should be noted that our meta-analytical approach may entail data validity problems, including biases due to subjectivity and different interpretations, and imprecise measurement. Furthermore, we encountered a shortage of data on less successful incubators, since reports tend to predominantly be published on the success stories of incubator development. We nevertheless included a substantial number of failing, or stagnating incubators (10) by using more than one data source and cross-checking data from different sources. As a result of our selection process, we were able to establish a set of 40 incubators. Growth of the number of new spin-offs (i.e., entrants) per year varies between –3.0 and +7.0, and follows a normal distribution. On the basis of the median (+1.25) and mean (+1.87), we divided the incubators into two classes, i.e., those experiencing relatively strong growth and those experiencing rather weak growth. In terms of growth, our selected incubators should represent a fair distribution of technology incubators in developed and in some emerging economies. However, the class of stationary incubators (zero-growth) may be somewhat underrepresented (for further details, see Soetanto and Geenhuizen, 2007).

In order to deal with the already mentioned difference in degree of the data's measurement accuracy (its 'fuzziness'), as well as with the sometimes low scale of measurement, we used a rough set analysis to assess the influence of the six factors identified above on incubator growth. Rough set analysis enables the transformation of an imprecise or incomplete ('fuzzy') collection of data, both quantitative and qualitative, into structured knowledge (Pawlak, 1991). Unlike conventional methods that are based on strict statistical assumptions, this analysis uses only one assumption, namely that the value of the "determining" factors can be categorized. In recent years, various studies have proved successful by using rough set as an analysis tool in different areas (Geenhuizen and Nijkamp, 2007; Goh and Law, 2003; Nijkamp et al., 2002a,b; Sanchis et al., 2007; Ubbels and Nijkamp, 2002; Warren et al., 2004).

[1] We selected countries with a comparable macro-economic growth between 1998 and 2002 (an average annual growth of 2.93%, with a maximum of 4.64% and a minimum of 1.10%) (OECD 2003; Web sites of National Statistical Offices). Two countries had a lower/higher growth rate, Japan (0.44%) and China (7.58%).

In rough set analysis, data is presented within an information table (IT), a matrix in which rows are labeled by *objects* and columns by *attributes*. Objects are arranged in an IT on the basis of their condition attributes (C) and decision attribute (D). These two attributes are analogous to the independent variables and the dependent variable in regression analysis. The condition attributes consist of the features describing the objects (the incubators), whereas the value of the decision attribute comprises concepts to be derived from the value of the condition attributes (growth of the incubators) (see Table 1). The basic procedure in rough set analysis is based on attribute reduction, i.e., finding a smaller set of attributes with the same, or similar, classification power as the original set of attributes. Two basic concepts in this context are *reduct and core*. A reduct is the essential part of an information table (subset of attributes) that distinguishes all objects discernible in the original information table. A core is a common part of all reducts. On the basis of a reduced information table, the procedure produces a number of *decision rules* by determining the decision attributes' value based on the condition of these values. A decision rule is presented in an "IF condition(s) ... THEN decision" format. Condition attributes situated in the so-called 'core' have the strongest explanatory power; they are indispensable in explaining the decision attribute's variation. All other condition attributes appear at a lower frequency rate within the decision rules. In general, a high frequency rate within the decision rules implies that the attributes stand out in a more pronounced way than others. In addition, the *coverage* (or strength) of the rules indicates the rate in which objects in a subset with the same decision attribute support the decision rule. In our study, the rough set estimations were conducted using ROSE2 software (Predki and Wilk, 1999).

Table 1. Information Table (selected cases for illustration purposes)

Object	Incubator region/city	Condition attributes (a)						Decision attribute (growth) (b)
		C1	C2	C3	C4	C5	C6	D
1	Trondheim, Norway	2	2	1	1	2	1	2
2	Laval, Quebec, Canada	2	1	1	2	1	1	1
3	Budapest, Hungary	2	1	2	1	1	2	1

(a) *Condition attributes:*
C1: Model of stakeholders' involvement (1: single stakeholder; 2: multiple stakeholder); C2: Level of urbanization (1: metropolitan areas; 2: non-metropolitan areas); C3: Uncertainty avoidance attitude (1: low; 2: high); C4: Model of support (1: conventional; 2: value-added); C5: Incubation strategy (1: pure research commercialization; 2: profit-focused); C6: Age of incubator (1: < 5 years; 2: ≥ 5 years) (in 2004).
(b) *Decision attribute:*
D: Average annual growth (1: relatively weak; 2: relatively strong)

Source: Adapted from Soetanto and Van Geenhuizen 2007.

We followed two steps in the application of rough set analysis. First, we reduced the base-population (40 incubators) to a random sample of 25 incubators, which facilitated the interpretation of the decision rules. To avoid sensitivity in the random sampling we worked with ten samples of this size. Secondly, we consistently used ROSE2 to measure the accuracy connected with the decision rules and, as an additional step, assessed the decision rules' prediction power, using samples previously not utilized (prediction accuracy). The accuracy of the core attributes and of the entire set of attributes turned out to be 75 and 76%, respectively, and the prediction accuracy amounted to a level between 60 and 80%, with 73% on average (for further details, see Soetanto and Geenhuizen, 2007). These levels justify the use of the decision rules in our interpretation of the rough set results.

5 Results of the Analysis

We employed two simple measures to assess the explanatory power of the rules, including (1) the frequency in which individual condition attributes appear within the strongest rules, and (2) the frequency in which the strongest rules appear among the ten samples. We conclude this analysis with the interpretation of various, typical strong rules.

We first analyzed the importance of individual condition attributes by using the strongest rules (rules with the highest 'coverage'). The label 'strongest rule' means that the rule is supported by the largest number of cases in the subset. Overall, based on our ten samples we identified the seven strongest rules associated with weak growth and the five strongest rules linked to strong growth (Table 2). For each condition attribute, the frequency of a strongest rule was divided by the total number of strongest rules in that category. For example, with regard to weak growth, the model of stakeholder involvement (C1) demonstrates a frequency rate of 43%.

Table 2. Frequency of Individual Factors in the Strongest Rules (a)

Weak growth (7 rules)	C1	C2	C3	C4	C5	C6
Frequency of appearance (%)	43.0 (3/7)	29.0 (2/7)	43.0 (3/7)	57.0 (4/7)	43.0 (3/7)	14.0 (1/7)
Strong growth (5 rules)	C1	C2	C3	C4	C5	C6
Frequency of appearance (%)	60.0 (3/5)	60.0 (3/5)	20.0 (1/5)	0	40.0 (2/5)	20.0 (1/5)

(a) C1-C6: The condition attributes (model of stakeholder involvement, level of urbanization, uncertainty-avoiding attitude, model of support, incubation strategy, and age of incubator).

On the basis of these outcomes, it is evident that only a few individual condition attributes have a strong explanatory power:

1. The model of conventional support (C4) for a weaker growth rate (57%). This indeed confirms our assumption that support that is merely tangible and fails to add social inputs, like training of individual start-ups, causes a negative impact.
2. Multiple stakeholder involvement (C1) for a stronger growth (60%). This also confirms our expectations that the involvement of different stakeholders introduces a favorable diversity of resources, including networks and knowledge flows.
3. A location in non-metropolitan areas (C2) (60%) for a stronger growth (60%). Contrary to the previous results, this particular outcome controverts our assumption of a stronger incubator development in metropolitan areas.

There may be two explanations for the result concerning the level of urbanization. First, in particular metropolitan areas, agglomeration economies may have turned into agglomeration diseconomies through road congestion, high land and real estate prices, labor market shortages, etc., whereas rural and peripheral areas are not particularly affected by such diseconomies, and, especially in Europe, benefit from regional development assistance provided by the EU and the national governments, thus lowering facility costs. Secondly, in agglomerated urban areas the incubation process may evolve spontaneously due to the availability of cheap and flexible rental space on the market, thereby reducing the supportive role of university incubators (redundancy). This may hold true for cities like Helsinki (Finland), Budapest (Hungary), and Atlanta (USA). In various large Scandinavian cities, it appeared that university incubators—given the existing supply in the market—offer some additional facilities at market prices. This, however, leads to the adverse effect of not being attractive for spin-offs and of consequently being underused. From a broader perspective, these cities do of course accommodate spin-offs and new economic growth. Table 2 indicates that the remaining three factors—uncertainty avoidance attitude, incubation strategy, and age of the incubator organization—are of minor importance in explaining growth patterns of incubators.

Next, we examined the frequency with which the strongest rules (top three for each sample) appeared in our ten samples (Table 3). Two strongest rules appeared in more than three samples, with one of the rules appearing in the majority of samples (six out of ten). These rules include (1) the combination of the model of (multiple) stakeholder involvement with the level of urbanization (non-metropolitan areas) (C1 & C2), and (2) the model of support provided to the start-ups (value-added support) coupled with incubation strategy type (a limited, profit-focused strategy) (C4 & C5).

Finally, we examined a typical sample of rough set results by focusing on the strongest rules (Table 4). It seems that the combination of the multiple stakeholder involvement model and location in a non-metropolitan area leads to a relatively strong growth of incubators, a rule applying to 6 out of 13 incubators (coverage of 46.2%). The second strongest rule is the combination of conventional support and an incubator strategy focusing on research commercialization; this combination leads to a relatively weak growth and applies to 5 out of 12 incubators (coverage of 41.7%). Equally strong, at least in the single sample displayed in Table 4, is the combination of a profit-focused strategy with a longer period of existence leading

Table 3. Frequency of the Strongest Rules (a)

Strongest rules (b)	Frequency rate (%)	Strong growth	Weak growth
C1 & C2	60.0 (6/10)	3	3
C4 & C5	40.0 (4/10)	2	2
C1 & C5	30.0 (3/10)	1	2
C1 & C6	20.0 (2/10)	0	2
C2 & C5	20.0 (2/10)	2	0

(a) Strongest rules that appear in more than one sample.
(b) C1-C6: the condition attributes (model of stakeholder involvement, level of urbanization, uncertainty avoidance attitude, model of support, incubation strategy, and age of incubator).

Table 4. Typically Strong Decision Rules (>40%) (one sample) (a)

Decision rules	Growth class	Strength (coverage) (%)
C1 = 2 and C2 = 2	Strong	46.2
C4 = 1 and C5 = 1	Weak	41.7
C5 = 2 and C6 = 2	Strong	41.7

(a) C1-C6 and their codes: see notes of previous tables.

to a strong growth. It must be noted that the level of coverage measured in rough set analysis usually does not exceed 50%.[2]

We may conclude that our findings suggest consistency in terms of the individual condition attributes and combinations of these in the strongest rules identifying the model of stakeholder involvement and level of urbanization as the two most significant determinants for the different growth patterns of incubators.

6 Discussion

Many governments perceive knowledge as the major engine of economic growth with entrepreneurship as a conduit for knowledge spillovers. However, investments in new knowledge and ideas may not automatically lead to spillovers and commercialization. Many hurdles must be overcome in innovative entrepreneurship, some of them originating in the institutional system and others influenced by entrepreneurial strength (or weakness). Due to the popularity of the knowledge-based economy, relevant policies have to date been implemented in many regions. Systematic knowledge, however, essential for constructive decision-making, is often absent. This is particularly true for incubators functioning as vehicles to enhance economic growth deriving from universities.

[2] We mention the following examples of maximum levels, i.e., 40% in Geenhuizen and Nijkamp (2007) (location-boundness of firms), 33.3% in Goh and Law (2003) (travel demand), 30% in Sanchis et al. (2007) (failure of firms), and 41.6% in Warren et al. (2004) (feasibility of waste management).

By adopting an institutional and evolutionary perspective, this chapter provides insight into incubators' performance and factors that determine their growth using a systematic, quantitative, and cross-national comparative approach. Two institutional factors consistently appeared as strong explanatory aspects, namely the model of stakeholder involvement and the nature of support provided to the incubated start-ups. For both factors, as well as a third albeit less important one, the 'direction' of influence appeared to be the same (as expected on the basis of evolutionary ideas): more modern, diversified developments tend to produce better results. Thus, diversity and hybridization manifesting themselves in a mix of stakeholders, a combination of conventional and added-value support (intended to improve entrepreneurs' business capabilities), and to a smaller extent, a certain shift towards a profit-focused strategy providing access to start-ups as "outsiders", all contribute to a stronger development. In addition, the level of urbanization of the incubator's location appeared to play a role. Incubators experiencing dynamic growth tend to be located in non-metropolitan (peripheral/rural) areas, while those experiencing weak growth are often found in large metropolitan areas. This unexpected pattern points to potential diseconomies of agglomeration in large cities, but also to potential redundancy of incubator initiatives there. Overall, the combination of multiple stakeholder involvement and a location in non-metropolitan areas turned out to be the most powerful "determinant" of dynamic incubator development.

The previous findings may be helpful in designing future policy decisions on technology incubators. This holds particularly true where incubator models reflecting diversity and hybridization produce more dynamic developments. However, it seems that relatively high levels of diversity and hybridization work contraproductively. In other words, there seems to be an optimum, but this optimum is not specified. Furthermore, the finding that technology incubators tend to be suboptimal in metropolitan areas also has policy implications. Instead of establishing new incubators in metropolitan areas, more efficient policy results may be achieved by supporting already existing incubation facilities and introducing higher levels of diversity and hybridization in established models.

This study was a first, broad inquiry into the pace of growth of technology incubators and, as such, subject to various limitations that may open doors to future research. Within the framework of this brief assessment, we were limited to modeling simple, one-way "determinant" relationships. Accordingly, we may have overlooked some of the complexity at hand, particularly regarding, e.g., feedback effects between incubator growth and stakeholder involvement, and effects from intervening factors and from non-linearity. For future research, a more advanced model needs to be developed and tested. Specifically, one of the limitations of this study is that the characteristics of the links between the university and the incubator, as well as of those with the start-ups, were not taken into account. Recent research indicates that the nature of university-start-up linkages has an impact on the speed of graduation and company failure (Rothaermel and Thursby, 2005), and accordingly, determines the number of incubated ventures per year. In addition, the growth of incubators in our study was only addressed in a quantitative way without information on important details about the new ventures involved. Future work should also include characteristics of incubated ventures, like their innovation level and market value, and their survival rate in the years after graduation, used as output indicators of incubators. Recently, Phan et al. (2005) described this point as the need to identify the relevant dependent variables that should be linked with the objectives of various incubators. In such a context, we refer to policy or strategy evaluation studies, which was not the nature of our particular research. A strict evaluation study seems realistic for those incubators with explicit and quantifiable objectives. Accordingly, broader and at the same time more in-depth research addressing this point should employ a cross-comparative and quantitative approach that requires the construction of extensive comparable databases, which consider a sufficient number of incubators, or smart ways of connecting existing data on a case-study basis. Thus, a large area lies before us, ready to be explored and deepen our knowledge of incubator performance. It is a fascinating quest in an age of continuing integration and hybridization of functions of universities, industry, and governments.

Acknowledgements This study was supported by the Delft Research Center Sustainable Urban Areas of Delft University of Technology.

References

Aernoudt R (2004) Incubators: Tools for entrepreneurship? Small Business Economics 23:127–135

Aghion P, Howitt P (1998) Endogenous growth theory. The MIT Press, Cambridge

Armstrong H, Taylor J (2001) Regional economics and policy. Blackwell, London

Audretsch D B (1998) Agglomeration and the location of innovative activity. Oxford Review of Economic Policy 14:18–29

Audretsch D B, Keilbach M C, Lehman E E (2006) Entrepreneurship and economic growth. Oxford University Press, Oxford

Autio E, Klofsten M (1998) A comparative study of two European business incubators. Journal of Small Business Management 36:30–43

Bergh J C J M van den, Button K J, Nijkamp P, Pepping G C (1998) Meta-analysis in environmental economics. Kluwer Academic Publishers, Dordrecht

Bergh J C J M van den, Stagl S (2003) Co-evolution of economic behaviour and institutions: Towards a theory of institutional change. Journal of Evolutionary Economics 13 030:289–317

Bøllingtoft A, Ulhøi J P (2005) The networked business incubator – Leveraging entrepreneurial agency? Journal of Business Venturing 20:265–290

Boucher G, Conway C, Meer E van der (2003) Tiers of engagement by universities in their region's development. Regional Studies 37(9):887–897

Castells M, Hall P (1994) Technopoles of the world. Routledge, London

Charles D R (2003) Universities and territorial development: Reshaping the regional role of English universities. Local Economy 18:7–20

Clarysse B, Wright M, Lockett A, Van de Velde E, Vohora A (2005) Spinning out new ventures: A typology of incubation strategies from European research institutions. Journal of Business Venturing 20:183–216

Cook T H, Cooper H, Cordray D S, Hartmann H, Hedges LV, Light R J, Louis T H, Mosteller F (1992) Meta-analysis for explanation. Sage, New York

Cooke P, Kaufmann D, Levin C, Wilson R (2006) The biosciences knowledge value chain and comparative incubation models. Journal of Technology Transfer 31:115–129

Druilhe C, Garnsey E (2004) Do academic spin-outs differ and does it matter? Journal of Technology Transfer 29:269–285

Duranton G, Puga D (2003) Microfoundations of urban agglomeration economies. In: Henderson J V et al (eds) Handbook of regional and urban economics, 4th edn. Elsevier, Amsterdam

Edquist C, Johnson B (1997) Institutions and organizations in systems of innovation. In: Edquist C (ed) Systems of innovation. Technology, institutions and organizations. Pinter, London

Etzkovitz H (2002) Incubation of incubators: innovation as a triple helix of university-industry-government networks. Science and Public Policy 29:115–128

Florida R (2002) The economic geography of talent. Annals of the Association of American Geographers 92:743–755

Freeman C, Perez C (1988) Structural crisis of adjustment, business cycles and investment behaviour. In: Dosi G, Freeman C, Nelson R, Silverberg G, Soete L (eds) Technical change and economic theory. Pinter, London

Geenhuizen M van, Nijkamp P (2007) Cities and footlooseness: In search of place-bound companies and effective location policies. Environment and Planning C: Government and Policy 25(5):692–708

Glaeser E L, Kolko J, Sainz A (2001) Consumer city. Journal of Economic Geography 1:27–50

Goh C, Law R (2003) Incorporating the rough sets theory into travel demand analysis. Tourism Management 24:511–517

Hackett S M, Dilts D M (2004) A systematic review of business incubation research. Journal of Technology Transfer 29:55–82

Hannon P D, Chaplin P (2003) Are incubators good for business? Understanding incubation practice – the challenges for policy. Environment and Planning C: Government and Policy 21:861–881

Hansen M T, Chesbrough H W, Nohria N, Sull D N (2000) Networked incubators. Hothouses of the new economy. Harvard Business Review 78(5):74–84

Hofstede G (1991) Culture and organizations: Software of the mind. British Library, London

Jacobs J (1961) The death and life of great American cities. Random House, New York

Leone R A, Struyck R (1976) The incubation hypothesis: Evidence from five SMAs. Urban Studies 13:325–331

Leydesdorff L (2003) A methodological perspective on the evaluation of the promotion of university-industry-government relations. Small Business Economics 20:201–204

Lockett A, Wright M, Franklin S (2003) Technology transfer and universities' spinout strategies. Small Business Economics 20:185–200

Maine E, Garnsey E (2006) Commercializing generic technology: The case of advanced materials ventures. Research Policy 35(3):375–393

McAdam M, Galbraith B, McAdam R, Humphreys P (2006) Business process and networks in university incubators: A review and a research agenda. Technology and Strategic Management 18(5):451–472

McKinnon S, Hayhow S (1998) State of business incubation industry. National Business Incubator Association (NBIA), Ohio

Mian S A (1997) Assessing and managing the university technology business incubator: An integrative framework. Journal of Business Venturing 12:251–285

Monck C S P, Porter R B, Quintas P, Storey D J, Wynarczyk P (1988) Science parks and the growth of high technology firms. Croom Helm, London

Morgan K (1997) The learning region: Institutions, innovation and regional renewal. Regional Studies 31:491–503

Mustar P, Renault M, Colombo M, Piva E, Fontes M, Lockett A, Wright M, Clarysse B, Moray N (2006) Conceptualizing the heterogeneity of research-based spinoffs: A multi-dimensional taxonomy. Research Policy 35(2):289–308

Nijkamp P, Rodenburg C A, Wagendonk A J (2002a) Success factors for sustaining urban brownfield development: A comparative case study approach to polluted sites. Ecological Economics 40:235–252

Nijkamp P, Burch M van der, Vindigni G (2002b) A comparative institutional evaluation of public-private partnerships in Dutch urban land-use and revitalization projects. Urban Studies 39:1865–1880

Nowotny H, Scott P, Gibbons M (2001) Re-thinking science. Knowledge and the public in an age of uncertainty. Polity Press, Cambridge

Pawlak Z (1991) Rough sets: Theoretical aspects and reasoning about data. Kluwer Academic, Dordrecht

Phan P H, Siegel D S, Wright M (2005) Science parks and incubators: observations, synthesis and future research. Journal of Business Venturing 20:165–182

Phillips R G (2002) Technology business incubators: How effective as technology transfer mechanisms? Technology in Society 24:299–316

Predki B, Wilk S (1999) Rough sets based data exploration using ROSE system. In: Ras Z W, Skowron A (eds) Foundation of intelligent systems. Lecture notes in artificial intelligence 1609. Springer, Berlin

Reid S, Garnsey E (1998) Incubation policy and resource provision: Meeting the needs of young innovative firms. In: Oakey R, During W (eds) New technology-based firms in the 1990s Vol. 5. Paul Chapman, London

Roberts E, Malone D (1996) Policies and structures for spin-off new companies from research and development organizations. R & D Management 26(1):17–48

Rothaermel F T, Thursby M (2005) Incubator firm failure or graduation? The role of university linkages. Research Policy 34:1076–1090

Rothaermel F T, Agung S D, Jiang L (2007) University entrepreneurship: A taxonomy of the literature. Industrial and Corporate Change 16(4):691–791

Sanchis A, Segovia M J, Gil J A, Heras A, Vilar J L (2007) Rough sets and the role of the monetary policy in financial stability (macroeconomic problem) and the prediction of insolvency in insurance sector (microeconomic problem). European Journal of Operational Research 181:1554–1573

Shane S (2004) Academic entrepreneurship. Edward Elgar, Cheltenham

Smilor R W, Gibson D V, Kozemtsky G (1988) Creating the technopolis: High technology development in Austin, Texas. Journal of Business Venturing 4:49–67

Soetanto D P, Geenhuizen M van (2007) Technology incubators and knowledge networks. A rough set approach in comparative project analysis. Environment & Planning B, Planning & Design 34:1011–1029

Ubbels B, Nijkamp P (2002) Unconventional funding of urban public transport. Transportation Research Part D 7:317–329

Varga A (1998) University research and regional innovation. A spatial econometric analysis of academic technology transfers. Springer, Heidelberg/Berlin

Viale R, Etzkowitz H (2005) Third academic revolution: Polyvalent knowledge: the “DNA” of the triple helix. Presentation at the 5th Triple Helix Conference, The Capitalization of Knowledge, May 18–21 2005, Fondazione Rosselli, Turino, Italy

Warren R H, Johnson J A, Huang G H (2004) Application of rough sets to environmental engineering models. In: Peters J F, Skowron A (eds) Transactions on rough sets I. Lecture Notes in Computer Science 3100:356–374

The Relationship between Universities of Professional Education and SMEs: Room for Improvement

Enno Masurel and Frans Werkhoven

Abstract Economic growth is highly dependent on innovation, both in the private and the public sector. The private sector consists of large firms on the one hand, and small- and medium-sized enterprises (SMEs) on the other. Much attention is presently being paid to accelerate the innovation process among SMEs. SMEs often lack the required knowledge to promote innovations directly. Given their strong regional focus and practical nature, universities of professional education make good partners for SMEs. This chapter presents new empirical evidence on this topic, based on a case study carried out in the Eastern part of the Netherlands. The results revealed that only a small number of the SMEs examined cooperated with public knowledge institutions, and when they did, mostly collaborated with universities of professional education. Problems in establishing and maintaining a relationship with knowledge institutions appeared to hinge on a lack of knowledge about available services and the required procedures, particularly among non-users. Users, on the other hand, had problems with the time-consuming nature of collaboration and ambiguous reference points. Accordingly, universities of professional education ought to follow a dual strategy to attract SMEs to the knowledge–interaction process. This strategy predominantly addresses firms that have not yet begun making use of such institutions. Communication plays a key role here. Once a relationship is established, cooperation must run smoothly. The most important requirement for establishing relations is an efficient organization, including a limited time impact, and clearly defined reference points.

Corresponding Author: Enno Masurel
VU Centre for Entrepreneurship, VU University Amsterdam, The Netherlands, e-mail: emasurel@feweb.vu.nl

Frans Werkhoven
Economic and Social Institute (ESI), Windesheim University of Professional Education, Zwolle, The Netherlands

M. V. Van Geenhuizen et al., (eds). *Technological Innovation Across Nations*,
DOI: 10.1007/978-3-642-00158-1_7,

1 Introduction

Innovation is a buzzword not only in the Netherlands, but in Western economic policy in general. In recent years, economic authorities have paid increased attention to small- and medium-sized enterprises (SMEs). Special consideration has been given to the relationship between SMEs and public knowledge institutions, particularly local universities of professional education. This chapter focuses on universities of professional education, in part because they—like SMEs—often play an important role within their geographic region, and thus stand in contrast to the more nationally and internationally-oriented 'traditional' universities and larger companies. Universities of professional education typically have a more practical approach, which complements SMEs' operational character. The term 'universities of professional education' is used here, though they are often also referred to as institutes of higher vocational education, higher-educational institutes, and polytechnics. This chapter intends to shed some light on SMEs' experiences with regional universities of professional education, as well as their views on these institutes. Accordingly, the primary problems SMEs face in the establishment and maintenance of relations with this particular public knowledge institution are addressed.

Our explorative research involves a case study that was carried out in the Eastern part of the Netherlands, namely, the Zwolle region and its surrounding area (Fig. 1). Since both SMEs and universities of professional education are represented in this region, it is a particularly suitable case study in this context.

We begin with an outline of SMEs' economic policy on innovation in the Netherlands, concentrating on the role of the National Innovation Platform. An assessment on the status of universities of professional education follows by providing insights derived from the literature about the contribution such universities make in terms of their curriculum (business program) and their cooperation with SMEs. Next, the fieldwork and empirical results on cooperation between SMEs and universities of professional education are presented. The chapter concludes with recommendations on ways to improve the relationship between public knowledge institutions (especially universities of professional education) and SMEs, from the viewpoint of entrepreneurs.

2 Setting the Scene

One of the focal points of current economic policy in the Netherlands is to encourage innovation by the private sector. The Innovation Platform was established in September 2003 (www.innovatieplatform.nl), since the Netherlands was not making the most of its economic and human potential. Values such as excellence, ambition, and entrepreneurship were reinvented. The policy of promoting innovation among the private sector was initiated following the 2000 European Lisbon Summit, at which EU leaders defined Europe's new mission: to become the most competitive and dynamic knowledge economy in the world. While Europe plays a leading role globally with regard to scientific excellence and the provision of highly-skilled human capital, it largely fails to convert science-based findings and inventions into wealth-generating innovations.

In reference to the Lisbon Treaty, the Dutch government declared its intention of making the Netherlands one of Europe's leading nations in the knowledge economy by 2010. Despite the government's numerous references to the Lisbon agenda, very little progress had actually been made by 2002. Meanwhile, the Dutch economy had been hit hard by economic recession, and several political lobbies advocated the establishment of a national coordination body, similar to the Finnish Science and Technology Policy Council (STPC) which develops and fosters innovation policy. Additional funds were made available for education, research, and innovation, and in August 2003, the Dutch Prime Minister launched the Innovation Platform along the lines of the Finnish STPC. The Innovation Platform's Implementation Office was set up in January 2004, and the Innovation Platform established by Royal Decree for a three-and-a-half year period, from January 2004 to July 1, 2007.

Finland's Science and Technology Policy Council had carried out research on entrepreneurial education involving local education authorities in Finnish municipalities, individuals responsible for business and industry-related issues, representatives of business enterprises, and members of local school boards in charge of basic education (Seikkula-Leino, 2007). The findings indicated that entrepreneurial education was increasing, but also that teachers lacked sufficient knowledge about the aims, contents, and work methods of entrepreneurial education, a circumstance that called for numerous improvements.

The Innovation Platform focused predominantly on large firms until summer 2005, when it introduced a series of measures targeting SMEs. One of the objectives was to bridge the gap between public knowledge institutions and SMEs, with the purpose of increasing innovation by SMEs. Universities of professional education play an important role here, because they—like SMEs—are more practically and regionally-oriented than traditional universities, and thus better complement the nature of SMEs. In this first round of measures, innovation performance contracts for entrepreneurs and universities of professional education were offered, and additional professors were appointed to teach courses that focused particularly on SMEs. So-called 'innovation vouchers' were introduced to encourage SMEs to invest in public knowledge institutions. In the first round, a total of 6,000 vouchers of €2,500 and €7,500 were made available to enterprises prepared to contribute a given amount to the subsidy voucher. This only paid off for the €7,500 vouchers which consisted of a €5,000 subsidy and a €2,500 contribution by the enterprise.

At the same time, the Organisation of Economic Co-operation and Development (OECD) stressed the importance of cooperation between the private sector and knowledge institutions. A number of self-evaluation studies in 14 regions across the world were commissioned by the OECD within the Institutional Management in Higher Education (IMHE) program, which ran from 2004 to 2007 (OECD, 2007a). Such initiatives are representative of the era of open innovation (Chesbrough, 2003). The model of open innovation generates the widespread dissemination of useful information. That is, the effects of innovation extend beyond the traditional boundaries of the firm, implying that the age of internally-oriented and centralized innovation within firms (closed innovation) has come to an end.

Building on these developments, this chapter examines the nature of the relationship between universities of professional education and SMEs, and the challenges perceived by SMEs in this regard. To derive further insights on this relationship, a case study was carried out in the Eastern part of the Netherlands. Based on the results of this explorative research, points of departure for future research can be formulated.

3 Universities of Professional Education Versus Traditional Universities in the Netherlands

Most of the literature on knowledge transfer focuses on universities.[1] A distinction is made between universities of professional education and traditional universities, the main difference being that universities of professional education do not offer Master's or doctoral programs and do not conduct fundamental research. Strictly speaking, a university of professional education can be defined as an institution offering instruction in a number of industrial arts and applied sciences, while a traditional university is a large and diverse institution of higher learning created to provide knowledge for life, prepare students for a profession, and to award degrees. As already mentioned, universities of professional education are more practically-oriented than traditional universities, and thus better complement the practical nature of SMEs. They also often play an important role within their geographic region—which also applies to SMEs—while traditional universities have a more national and international orientation. Consequently, universities of professional education may be more suitable partners than traditional universities to meet the specific needs of SMEs. In the Netherlands, there is still a clear and formal distinction between traditional universities and universities of professional education in contrast to other countries, in which universities of professional education have already been accredited. However, with the introduction of the Bachelor/Master's system in higher education in the Netherlands, the gap between traditional universities and universities of professional education has narrowed, a development which may continue in the future.

Several scholars have pointed out that SMEs generate innovations differently than larger firms, emphasizing that 'a small business is not a little big business' (Welsh and White, 1981). Therefore, it is important to take the specific characteristics underlying SMEs' innovation process into account. Size is not the only factor that makes SMEs unique; their regional orientation and practical nature are significant characteristics as well (Thurik, 1996; Nooteboom, 1994; Yap and Shouder,

[1] See, for example, Rosenberg and Nelson (1994); Lee (1996); Harmon et al. (1997); Argyres and Liebeskind (1998); Stephan and Everhart (1998); Bercovitz et al. (2001); Nelson (2001); Stefan (2001); Acs et al. (2002).

1994). Bashkaran (2006) furthermore asserts that incremental innovations (i.e., ongoing improvements to products and processes) are essential for SMEs.

4 Public Knowledge Institutions and SMEs

The more regional and practical approach applied by universities of professional education cultivates their strong focus on SMEs and more frequent and direct contacts with them. SMEs' experiences with and opinions on regional universities of professional education are examined here to emphasize the importance of entrepreneurial education programs that are targeted at inducing the desirability and feasibility among students to start their own business (Peterman and Kennedy, 2003). However, if such educational programs are to be effective, a thorough understanding of the needs and demands of SMEs is necessary.

According to Ryans et al. (1987), the relevance of small businesses in the USA and Canada is not reflected in a typical business school's curriculum:[2] most courses are directed toward preparing students for careers in Fortune 500-type companies, large non-profit organizations, or the public sector. However, Ryans et al. (1987) argue that certain aspects of small business management could easily be integrated into the general business curriculum, while others are unique to small businesses and deserve special attention in a separate course on small business management. Such courses should include venture capital, entrepreneurship, business organization, business planning, government relations, and opportunities for women and minorities. It is also important to keep small business courses up-to-date by regularly inviting external guest speakers with special expertise.

Vesper and Gartner (1997) found that the most frequently offered business courses at both the undergraduate and graduate levels include entrepreneurship or the establishment of new firms, small business management, field projects/venture consulting, starting and running a firm, venture plan writing, and venture finance. The top seven criteria to rank entrepreneurial programs as suggested by Vesper and Gartner (1997) include: (i) number of different courses offered, class size, number of credits or class sessions, methods of teaching, the instructors' credentials, etc.; (ii) faculty publications, such as books, journal and magazine articles, number of citations, the target audience, length of articles, etc.; (iii) impact on community, like public symposia, student consulting projects, company spin-offs; (iv) accomplishments by alumni, e.g., number of start-ups by alumni, alumni participation as investors, partners, employees, etc. in other ventures; (v) innovations contributed by alumni to the entrepreneurial programs or innovations within the programs at institutions of higher education; (vi) achievements by alumni start-ups; and (vii) reaching out to scholars through the organization of conferences, sponsoring journals, creating and distributing new teaching materials, publishing newsletters, and supporting different journals.

[2] Business schools refer to graduate schools offering business and business-related courses and MBA programs.

Corsten (1987) was one of the first to research university–SME cooperation in the Netherlands. He surveyed SMEs on their experiences in collaborating with universities, particularly with regard to technical and/or scientific problems. Approximately one fourth of the respondents had cooperated with universities in the last ten years, especially larger SMEs. The majority of contacts were established through alumni who were now working for the given enterprise, through personal contacts with scientists (for example, contacts made at conferences or seminars), as well as through transfer agencies. In general, SMEs were satisfied with their experiences with university collaboration. Corsten (1987) differentiates between SMEs that have worked together with universities before and those that have not. The most pertinent problem for both these groups related to obstacles in communication. The lack of financial resources was only of moderate significance for both groups, whereas the lack of practicality revealed considerable divergence between the two groups: those who had previously worked with universities attached less value to this aspect than those that had not. This underscores the importance of personal communication, in particular personal interaction and trust. These issues must be addressed, taking the given enterprises' interests into consideration and keeping cooperation flexible.

Birdthistle (2006) found that third-level educational institutes in Ireland are not really used for the purpose of marketing research, R&D, and consultancy. This coincides with our findings that the current education system ought to have a more entrepreneurial focus and be more supportive of entrepreneurs. Instructors of business courses are often not involved in increasing entrepreneurs' awareness of the services that institutions of higher education can provide, despite the fact that relationships between the two can take many forms (Billet and Seddon, 2004; McCarthy et al., 1997).

5 Data Collection

The empirical research was conducted in the region administered by the Chamber of Commerce of Zwolle and its surrounding area in the Eastern part of the Netherlands (Fig. 1). Although the case study was carried out in the area of Zwolle, the results may apply to other regions and countries as well. In terms of firm size, the Zwolle region is to some degree representative of the Netherlands as a whole—close to 90% of the firms employ less than ten people. One notable difference is that the service sector plays a less important role in the Zwolle region than it does in the Netherlands as a whole.

According to an ING Bank study from 2004 in the IJsselland region, which is to some degree comparable to the Zwolle area, innovation potential rests on four cornerstones: the private sector (i.e., its efforts to develop innovations); labor potential (the degree of human capital in the innovation process); educational and knowledge infrastructure (institutes that transfer and create knowledge and generate an educated workforce); and other environmental factors (such as incubators for new ventures, available space, non-profit-sector, and quality of the environment).

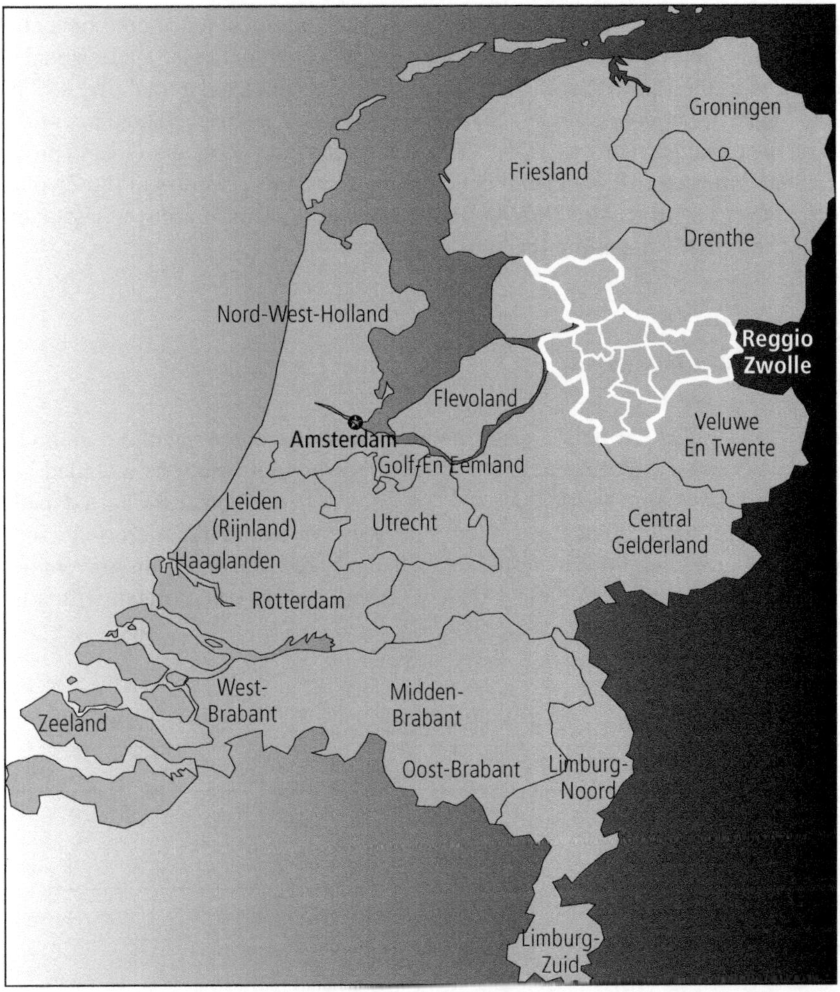

Fig. 1. The Zwolle Region in the Netherlands

Interestingly, the ING Bank (2004) study concluded that the IJsselland region's innovation potential was weaker than that in the rest of the Netherlands. The score for the private sector and labor potential was lower, while the region's educational and knowledge infrastructure, as well as the score for other environmental factors was higher than the overall results for the Netherlands. The reason for the Netherlands' higher performance with regard to innovation potential is that the private sector and labor potential are assigned more weight than the other two cornerstones.

For our study, we looked at the services, manufacturing, construction, and wholesaling sectors, in the assumption that firms from these sectors are more interested than others in establishing relations with knowledge institutions. We did not limit our research to technology-driven ventures, but also included more conventional firms, as they are particularly suitable partners for universities of professional education due to their practical and regional focus. We contacted a total of 1,292 firms with at least 5, but not more than 50 employees. 401 firms participated in our study, that is, the response rate was 31.0%. This is a satisfactory score overall and provides sufficient data to arrive at general conclusions about these sectors in the Zwolle region. In terms of firm size, 38.1% of the participating firms employed less than ten people, and 61.9% had more than ten employees.

6 Empirical Results

Table 1 illustrates that only a small number of SMEs in the Zwolle region (13%) have collaborated with one or more of the knowledge institutions included in our study (traditional universities, as well as universities of professional education). Medium-sized enterprises cooperated with these institutes far more than smaller enterprises did. Over 90% of respondents had worked together with universities of professional education, while less than 10% had already collaborated with a traditional university.[3]

Table 1. Cooperation with One of the Specified Knowledge Institutions

	Small firms	Medium-sized firms	All firms
Yes	5.9	17.4	13.0
No	94.1	82.6	87.0
Total	100	100	100

N= 399
P-value: 00

Research carried out by the OECD revealed that less than 10% of SMEs in Europe have collaborated with both higher education and government institutions (OECD, 2007b,c). The percentage of Finnish SMEs that collaborate with knowledge institutions was highest with 10%, whereas the result for the Netherlands was only 3%. Besides Finland, Sweden and Denmark's SMEs also exhibited higher levels of cooperation with knowledge institutions than the Netherlands, while some Southern and Eastern European countries, like Italy and Spain, or Poland and Hungary, had lower scores.

The OECD score for the Netherlands is considerably lower than the level of cooperation in the Zwolle region (13%). This divergence may possibly be linked to the different definitions of collaboration. In the OECD studies, 'collaboration'

[3] We included ten higher education institutions located in the Eastern part of the Netherlands in our study. Our approach was inspired by two characteristics that are typical of SMEs, their strong regional focus and their practical nature.

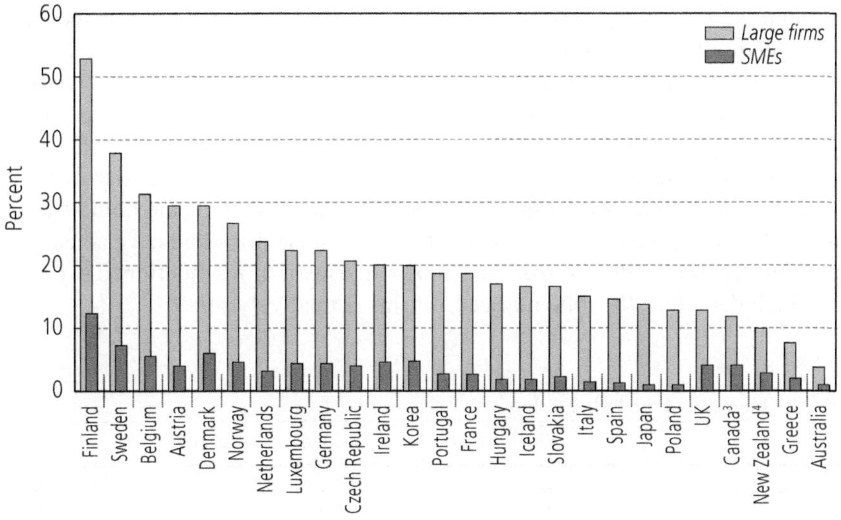

Fig. 2. Firms Collaborating with Higher Education Institutions, by Size[1], 2002–2004[2] (Source: OECD, 2007c)

1. SMEs: 10–249 employees for European countries, Australia and Japan (persons employed); 10–99 for New Zealand, 10–299 for Korea, 20–249 for Canada.
2. Or nearest available years.
3. Manufacturing sector only.
4. Refers to firms that cooperate with Crown Research Institutes or other research institutes.

Statlink http://dx.doi.org/10.1787/11766272755

involves active participation in joint innovation projects. Our questionnaire, on the other hand, inquired whether the firms 'made use' of knowledge institutions, which covers a broader scope in terms of relations. Working with this definition and a specification of the region's knowledge institutions seemed more appropriate with regard to the practical nature of SMEs.

The OECD reports also show that size plays a role: large firms were four times more likely to collaborate with knowledge institutions than SMEs. This result is in line with our findings about the effect of size (i.e., that medium-sized firms collaborate more frequently with institutions of higher learning than smaller ones).

Several studies have examined the cooperation between higher education institutions and the private sector. For example, a survey of the UK's manufacturing industry indicated that 27% of the firms cooperated with knowledge institutions (Laursen and Salter, 2006). Though the difference between our results and those of the UK's survey can be attributed to the inclusion of large firms in the UK's manufacturing segment as well, the difference can possibly be also explained by the slightly stronger innovative character of firms in the UK as compared to those in the Zwolle region.

In terms of intensity of the relationship, nearly half of the respondents who had cooperated with one or more of the institutions included in our study claimed that the intensity of the relationship was quite low (Table 2). Almost one quarter described the relationship as being somewhat intense. The relationship was perceived as being very intense by 14.9%, and finally, 12.8% claimed that the intensity of the relationship fluctuated over time. The assessments on the intensity of the relationship with knowledge institutions by both the small- and the medium-sized firms diverged only slightly. It must be noted that only those firms that have actually been involved with one of the ten specified knowledge institutions mentioned in our study were included, thus, the total number of firms is relatively small.

Table 2. Intensity of the Relationship

	Small firms	Medium-sized firms	All firms
Low intensity	42.9	50.0	48.9
Moderate intensity	28.6	22.5	23.4
High intensity	14.3	15.0	14.9
Fluctuating intensity	14.3	12.5	12.8
Total	100	100	100

N= 47
P-value: 0.98

Table 3 indicates that entrepreneurs usually initiate cooperation rather than the higher education institutions. Occasionally, a third party (e.g., a consultancy firm) may also be involved in establishing a relationship between an enterprise and a knowledge institution. Alternatively, relationships can emerge by coincidence, through alumni working for a given firm, etc. There are no significant differences in the way small- and medium-sized enterprises establish relationships with knowledge institutions.

Table 3. Ways of Establishing Collaboration

	Small firms	Medium-sized firms	All firms	P-value
Initiated by the entrepreneur	57.1	37.5	40.4	0.33
Initiated by the institution	28.6	27.5	27.7	0.95
Through a third party	0.0	17.5	14.9	0.23
Coincidence	14.3	12.5	12.8	0.90
Through alumni	0.0	12.5	10.6	0.32
Other	0.0	15.0	12.8	0.27

N= 47

With regard to the type of cooperation, more than half are internships (Table 4). Nearly a quarter of all collaborations between SMEs and knowledge institutions involve training and development, followed by joint innovation ventures. Contracting students, contract research, and alumni outreach are additional types of collaboration that are more common. The category 'other' includes informational meetings, the joint use of equipment, etc. Table 4 reveals that the number and types of cooperation with knowledge institutions established by both small- and medium-sized enterprises differ only slightly.

Table 4. Types of Collaboration

	Small firms	Medium-sized firms	All firms	P-value
Internships	42.9	60.0	57.4	0.39
Training & development	28.6	22.5	23.4	0.73
Joint innovation projects	14.3	12.5	12.8	0.90
Contract students	0.0	12.5	10.6	0.32
Contract research	14.3	2.5	4.3	0.15
Alumni outreach	0.0	5.0	4.3	0.55
Other types of collaboration	0.0	20.0	17.0	0.19

N= 47

In addition, the reasons for establishing relationships with knowledge institutions also do not differ significantly between smaller and medium-sized firms, as illustrated in Table 5. Marketing and quality are the most frequent reasons for collaborating, representing nearly one third of all cooperations, followed by strategy and finance. Work conditions and environment barely play a role, and export plays none at all. The category 'other' includes product development, technical assistance, etc.

Table 5. Reason for Collaboration

	Small firms	Medium-sized firms	All firms	P-value
Marketing	42.9	26.7	31.8	0.54
Quality	50.0	23.3	31.8	0.20
Strategy	28.6	16.7	20.5	0.57
Finance	14.3	13.3	13.6	0.87
Work conditions	7.1	10.0	9.1	0.62
Environment	0.0	3.3	2.3	0.44
Export	0.0	0.0	0.0	–
Other	35.7	26.7	29.5	0.85

N= 44

The most important reason for cooperating with a higher education institution is quality improvement, which applied to more than half of the respondents (Table 6). Over one third of the respondents had pursued a partnership with knowledge institutions to increase their efficiency, followed by the intent to improve sustainability and to facilitate the development of innovations. The underlying reasons for the establishment of such relations hardly differ between small- and medium-sized firms.

Table 6. Purpose of the Collaboration

	Small firms	Medium-sized firms	All firms	P-value
Quality improvement	42.9	54.1	52.3	0.59
Efficiency increase	42.9	37.8	38.6	0.80
Sustainability contribution	14.3	27.0	25.0	0.48
Facilitating innovations	28.6	13.5	15.9	0.32
Other	14.3	27.0	25.0	0.48

N= 44

Table 7 indicates that the majority of respondents that collaborate with knowledge institutions in the Zwolle region are satisfied—and in fact, even very satisfied—with the relationship. Very few firms reported being dissatisfied. A minority is neither satisfied nor dissatisfied, and a few respondents' satisfaction level has fluctuated. There were virtually no differences between small and medium-sized businesses in terms of levels of satisfaction or dissatisfaction. We presume that realistic expectations by the SMEs and good performance by the knowledge institutions are the reason for this satisfaction. It must also be noted here that only very few enterprises (7%) received a government subsidy which only partially covered the expenses of the collaboration, implying that this factor hardly had any influence on the level of satisfaction.

Table 7. Satisfaction with the Relationship with Knowledge Institutions

	Small firms	Medium-sized firms	All firms
Very dissatisfied	0.0	2.6	2.2
Dissatisfied	0.0	2.6	2.2
Neither	14.3	10.3	10.9
Satisfied	71.4	71.8	71.7
Very satisfied	14.3	10.3	10.9
Fluctuating	0.0	2.6	2.2
Total	100	100	100

N= 46
P-value: 98

Finally, the problems perceived by SMEs in establishing and maintaining relationships with knowledge institutions were analyzed. First, it is worth mentioning that over one third of the respondents had not experienced any problems (Table 8). The most often stated problem was the lack of knowledge about the services offered by the institutions (37%). The belief that knowledge institutions served education, not business development, was widespread. A smaller share of respondents was not familiar with the required procedures, e.g., issues relating to delivery time and confidentiality (30%). The third most mentioned problem was that cooperation is too time-consuming, an issue that preoccupies medium-sized businesses much more than it does small ones. The problem of finding a suitable individual (partner) within the institution ranked fourth and medium-sized firms perceived this to be more problematic than smaller enterprises. Communication problems were also mentioned slightly more often by medium-sized businesses than by smaller firms.

Table 8. Problems Perceived in the Collaboration with Knowledge Institutions

	Small firms	Medium-sized firms	All firms	P-value
Available services unknown	35.0	37.4	36.5	0.63
Procedures unknown	27.1	31.1	29.6	0.42
Too time-consuming	7.9	14.9	12.3	0.04
Not able to find suitable individual (partner) within the institution	5.0	11.1	8.8	0.05
Too expensive	6.4	3.0	4.3	0.11
Unclear references	2.9	3.4	3.2	0.77
Lack of understanding	0.7	3.4	2.4	0.10
Confidentiality/secrecy	0.0	0.4	0.3	0.44
Other problems	12.9	10.6	11.5	0.51
No problems	38.6	34.0	35.7	0.38

N=375

We also examined the differences between firms that have cooperated with one or more universities of professional education and firms that have not (Table 9). Firms that collaborate with knowledge institutions perceive fewer problems with such partnerships than do firms that have not yet done so, a result that corroborates Corsten's findings (1987). Firms that had not cooperated with knowledge institutions were much less aware of the available services offered by these and far more unfamiliar with the procedures involved. Evidently, experience plays a role here, since firms that have worked with knowledge institutions perceive fewer problems than firms with no experience. That is, lack of knowledge about available services and the related procedures may be reasons why non-collaborative firms have not yet pursued any partnerships with knowledge institutions. Surprisingly, two issues were considered more problematic by collaborative firms than by the non-collaborative firms surveyed, namely that maintaining the relationship was time-consuming and that recommendations provided by the institutions were difficult to understand. Previous experience by both parties involved in a collaborative project may play a role in this context, or perhaps firms that cooperate with knowledge institutions take such problems for granted. Other issues such as difficulty in finding a suitable individual (partner) within the institution, excessive expenses, a lack of understanding between the firm and institution, etc. were to a large degree shared equally by both collaborative and non-collaborative firms. Confidentiality only played a minor role.

Table 9. Problems Perceived in Establishing and Maintaining a Relationship with Knowledge Institutions

	Users	Non-users	All firms	P-value
Available services unknown	23.5	39.0	36.9	0.03
Procedures unknown	19.6	31.3	29.7	0.09
Too time-consuming	19.6	11.0	12.2	0.08
Not able to find suitable individual (partner) within the institution	15.7	7.7	8.8	0.10
Too expensive	3.9	4.3	4.2	1.00
Unclear references	7.8	2.5	3.2	0.06
Lack of understanding	2.0	2.5	2.4	1.00
Confidentiality/secrecy	2.0	0.4	0.3	0.14
Other problems	9.8	11.7	11.4	0.70
No problems	49.0	33.4	35.5	0.03

N=375

7 Conclusion and Recommendations

Our research shows that 13.0% of the SMEs in the region bordering the Randstad conurbation in the Netherlands have interacted with public knowledge institutions. This percentage is considerably higher than that recorded in comparable studies carried out by the OECD. One explanation may be our definition of collaboration, which involves a broader set of relationships, as well as our focus on specific higher education institutions in the Zwolle region. The results also indicate that size plays a role: medium-sized firms engage in collaborative partnerships with knowledge institutions far more often than small firms. Most collaborations between universities of professional education and SMEs involve internships and other opportunities for students. SMEs engage in such relations for a number of reasons: to gain access to new technologies (knowledge transfer), to obtain affordable consultation/advice and hands-on experience, or for reasons of social responsibility. The partnerships with knowledge institutions are generally weak, which applies to both small- and medium-sized firms. This also holds true for how the cooperation came about (usually upon the entrepreneur's initiation), the type of cooperation (mostly internships), reasons underlying the cooperation (predominantly marketing and quality), the purpose of cooperation (quality improvement), and satisfaction with the relationship (mostly satisfactory).

The low level of cooperation prompted an in-depth study on the problems perceived by firms with regard to collaboration with knowledge institutions. Frequently perceived problems included a lack of awareness about services offered by knowledge institutions and unfamiliarity with the related procedures. However, over one third of the respondents had not experienced any problems. Our analysis also revealed that medium-sized firms more so than small ones had faced problems relating to the time-consuming nature of such relations, difficulty in finding a suitable individual (partner) within the institution, and misunderstandings within the relationship. Furthermore, we observed some differences between firms that had already cooperated with knowledge institutions and those that had not. Firms with no experience in collaboration lacked knowledge about available services and procedures more often, which seems to be the primary reason for not pursuing a partnership with a knowledge institution. Firms that have cooperated with such institutions and are familiar with the procedures involved reported problems with regard to the time-consuming nature of this form of interaction and with clearly understanding the references provided by the institution.

Based on these findings, several recommendations can be formulated. Future research in other regions or countries can benefit from our approach of examining the level of cooperation between SMEs and knowledge institutions. Knowledge institutions ought to implement a two-way strategy to attract more SMEs for the purpose of exchanging knowledge and enhancing learning processes. One strategy should address those firms that do not collaborate with knowledge institutions. Since a lack of familiarity with available services and associated procedures is the primary reason for SMEs to not pursue a partnership with knowledge institutions, communication and raising awareness are key for improving this situation. It may be useful to involve stakeholders like the Chamber of Commerce and local business associations, since they may be able to convince entrepreneurs of the benefits of participating in an interactive learning process with knowledge institutions. However, once a relationship is established, cooperation must run smoothly; therefore, the main complaints by firms that have already cooperated with knowledge institutions need to be taken seriously, even more so because these problems could be grounds for ending all cooperation with universities in general. The most important conclusion our case study draws is that efficient coordination (limiting time impact, providing clear instructions) is key for successful collaboration with knowledge institutions. Stakeholders may play an important role in the coordination of such partnerships.

Acknowledgements The data analysis by Jan Holleman, assistance with the text by Patricia van Hemert, and helpful comments by the first editor and two anonymous reviewers are gratefully acknowledged. This research project was made possible by the Chamber of Commerce for the Zwolle region.

References

Acs Z J, FitzRoy F R, Smith I (2002) High technology employment and knowledge spillovers. In: Acs Z J, de Groot H L F, Nijkamp P, The emergence of the knowledge economy, Springer, 155–172

Argyres N S, Liebeskind J P (1998) Privatizing the intellectual commons: Universities and the commercialization of biotechnology. Journal of Economic Behavior and Organization 35(4):427–454

Bashkaran S (2006) Incremental innovation and business performance: Small and medium-size food enterprises in a concentrated industry environment. Journal of

Small Business Management 44(1):64–80

Bercovitz J, Feldman M, Feller I, Burton R (2001) Organizational structure as a determinant of academic patent and licensing behavior: An exploratory study of Duke, Johns Hopkins and Pennsylvania State Universities. Journal of Technology Transfer 26(1-2):21–35

Billet S, Seddon T (2004) Building communities through social partnerships around vocational education and training. Journal of Vocational Education and Training 56(1):51–68

Birdthistle N (2006) Entrepreneur's perception of tertiary level educators and public sector policy for new venture creation in the mid-West region of Ireland. Irish Business Journal 2(1):98–105

Chesbrough H W (2003) The era of open innovation. Sloan Management Review 44(3):35–41

Corsten H (1987) Technology transfer from universities to small and medium-sized enterprises: An empirical survey from the standpoint of such firms. Technovation 6(1):57–68

Harmon B, Ardishvili A, Cardozo R, Elder T, Leuthold J, Parshall J, Raghian M, Smith D (1997) Mapping the university technology transfer process. Journal of Business Venturing 12(6):423–434

ING Bank (2004) Ondernemen in IJsselland, regio in economisch perspectief ('Enterprising in IJsselland, region in economic perspective'). ING Bank, Amsterdam

Laursen K, Salter A (2006) Open innovation: The role of openness in explaining innovation performance among UK manufacturing firms. Strategic Management Journal 27:131–150

Lee Y S (1996) 'Technology transfer' and the research university: A search for the boundaries of university-industry collaboration. Research Policy 25(6):843–863

McCarthy A, Morris P J, Winn J (1997) A new look at undergraduate entrepreneurship education. Paper presented at the International Council for Small Business Conference, San Francisco, California, 21–24 June, 1997

Nelson R R (2001) Observations on the post-Bayh-Dole rise of patenting at American universities. Journal of Technology Transfer 26(1-2):13–19

Nooteboom B (1994) Innovation and diffusion in small firms: Theory and evidence. Small Business Economics 6(5):327–347

OECD (2007a) `http://www.obs-pascal.com/node/524(or766)`

OECD (2007b) Policy brief, November 2007, Innovation and Growth, OECD, Paris

OECD (2007c) OECD science, technology and industry scoreboard 2007. Innovation Policy, C–5. Collaboration with public research organizations by innovating firms. `http://www.sourceoecd.org/scoreboard`

Peterman N R, Kennedy J (2003) Enterprise education: Influencing students' perceptions of entrepreneurship. Entrepreneurship Theory & Practice 28(2):129–144

Rosenberg N, Nelson R R (1994) American industries and technical advance in industry. Research Policy 23(3):323–348

Ryans C C, Ryans Jr. J K, Pearson M A (1987) Educators' views on small business training. Journal of Small Business Management 27 (3):80–83

Seikkula-Leino J (2007) Curriculum reform and entrepreneurship education. Publications of the Finnish Ministry of Education 2007:28, Helsinki

Stefan P E (2001) Educational implications of university-industry technology transfer. Journal of Technology Transfer 26(3):199–205

Stephan P E, Everhart S S (1998) The changing reward to science: The case of biotechnology. Small Business Economics 10(2):141–151

Thurik A R (1996) Innovation and small business. Small Business Economics 8(3):175–176

Vesper K H, Gartner W B (1997) Measuring progress in entrepreneurship education. Journal of Business Venturing 12 (5):403–421

Welsh J A, White J F (1981) A small business is not a little big business. Harvard Business Review July-August: 18–32

Yap C M, Shouder W E (1994) Factors influencing new product success and failure in small entrepreneurial high technology electronic firms. Journal of Product Innovation Management 11(5):418–432

Small Biotechnology Industry and Institutional Transformation: A Case Study of The Netherlands

Marina van Geenhuizen

Abstract Many governments have adopted policies to enhance the establishment and growth of new biotechnology ventures in an attempt to benefit from the commercialization of new research findings in biotechnology. This chapter examines the policy of the Dutch government in this regard. Institutional shortcomings are identified and the policies implemented to counter these shortcomings evaluated. Furthermore, the 'landscape' of biotechnology in the Netherlands is analyzed by examining whether a biotech firm's particular location within a cluster contributes to innovation. The findings, together with an evaluation of the actual changes in the industry and within the relevant institutions, provide input into a critical reflection of the government's policy. Developments in the Netherlands indicate that academic entrepreneurship can increase substantially within a few years, if a comprehensive approach is implemented. However, the lacking involvement of large domestic companies, the institutional hurdles for financing start-up companies' subsequent growth stages, as well as the impacts of restrictive rules and legislation are challenges that are quite difficult to overcome and costly in terms of time.

Key words: Biotechnology, New Firm Formation, Policy, Clusters, Institutional Transformation, the Netherlands

1 New Business Opportunities and Institutional Constraints

The extent to which companies may utilize new knowledge depends not only on the availability of new knowledge sources, but on various institutions that mediate knowledge transfer, as well as barriers obstructing such transfer (Audretsch et al., 2006). The Netherlands, as a small economy, produces a considerable amount

Marina van Geenhuizen
Faculty of Technology, Policy and Management, Delft University of Technology, PO Box 5015, 2600 GA Delft, The Netherlands, e-mail: m.s.vangeenhuizen@tudelft.nl

M. V. Van Geenhuizen et al., (eds). *Technological Innovation Across Nations*,
DOI: 10.1007/978-3-642-00158-1_8,

of scientific knowledge, particularly in areas that lend themselves to commercialization. Yet the country lags behind in the economic utilization of the knowledge produced. Though this so-called "knowledge paradox" is also evident in other Western European countries, it is particularly strong in the Netherlands. If we take the number of spin-offs from public research institutes as an indicator of knowledge commercialization, the Netherlands falls behind benchmark countries by 30 to 40% (MEA, 2003a).[1] The knowledge paradox holds true especially for the biotechnology sector.

Modern biotechnology—also referred to more broadly as life sciences—is an enabling technology that can be applied in a wide range of sectors, e.g., in the pharmaceutical and medical sectors, the agriculture and food industry, and in the environmental sector. It comprises technologies linked to recombinant DNA techniques and cell fusion and, more recently, fields such as combinatorial chemistry and genome analysis. The OECD (2006) defines biotechnology quite broadly as "the application of science and technology to living organisms, as well as parts, products and models thereof, to alter living or nonliving materials for the production of knowledge, goods and services" (p. 7). Currently, the most important technological trend is genomics, including, among others, transcriptonomics (the functioning of genes), metabolomics (metabolism in the cell), and systems biology (MEA, 2003b). Results from new genomics research combined with developments in existing technology, for example, cloning and cell and tissue culturing, indicate the emergence of drastically new approaches for identifying and curing diseases, which may potentially have a *disruptive* impact on the present organization of the medical field and its related sectors. Furthermore, biotechnology in health care has increasingly expanded to involve three other generic technologies, namely informatics and computational science (e.g., in terms of data-mining to identify new effective compounds and the processing of genomics data), remote diagnostics and clinical trials, new materials technology (e.g., for the creation of new types of artificial bone and tissue), and nanotechnology (e.g., applied to bio-sensors). Thus, a wide variety of science and engineering fields are playing an important role in the quest for innovations. Additionally, market demand and institutions, such as regulations for new drugs or new food ingredients, intellectual property rules, and national health and safety control systems have an effect on the innovation system (Malerba, 2006; McKelvey et al., 2004).

The combination and integration of the above-mentioned technologies opens doors to entirely new applications for products, processes, and services (in terms of scale and functions), and provides a broad range of new business opportunities for innovative companies. However, such innovative companies also face a higher degree of uncertainty. These uncertainties, which are much more pronounced in the biotech than in any other high technology sector, may jeopardize a firm's survival (Audretsch, 2001; van Geenhuizen, 2003). Risks include the failure of research, for example, a delayed discovery of harmful side effects in new drugs following a long

[1] In the Netherlands, the number of spin-offs relative to turnover is 1.86% (3.09% in benchmark countries); relative to employment at knowledge institutes, the figure is 1.69% (versus 2.41% in benchmark countries).

and intensive period of development and testing. Failure may also occur at an earlier point in time, for example, when venture capital is not achieved after the initial rounds of seed capital.

Within the context of risk-taking strategies adopted by biotechnology companies, it is important to distinguish between different business models, i.e., between product, tool, and service models. Product companies are highly innovative and invest heavily in R&D to develop and launch new products in the market. Tool companies focus on developing platform technologies, such as breeding transgenic animals for the production of specific pharmaceutical substances, and usually generate revenues through out-licensing and royalties. In contrast, service companies carry out contracted research, or offer standard or customized services to product and tool companies, thus generating revenue from the onset of the venture. Finally, hybrid companies typically combine activities attributed to the other models described, e.g., combining routine contract research with R&D, thereby enabling the financing of R&D internally (Walsh, 1993). Opting for the hybrid model can be classified as a risk-aversive strategy. Another popular strategy to reduce risk is collaboration with large pharmaceutical companies. Entering cooperative alliances with large pharmaceutical companies serves several purposes, namely, to improve a small venture's financial position, gain experience in testing and trial procedures, and access to marketing and global distribution channels. Significant and positive side effects may include an increase in credibility, which is beneficial for raising further funds, as well as for the development of new collaborations (Powell, 1998; Senker and Sharp, 1997). More recently, an increase of mergers (acquisitions) and alliances within the EU's biotechnology sector has become evident, implying that innovations increasingly derive from various sources and skills represented in numerous and diverse collaborative networks (Salman and Saives, 2005).

Forecasting studies anticipate that biotechnology will foster tremendous growth in a number of sectors. Sectors in which most of the new technologies are delivered by biotechnology companies are estimated to reach a total world market share of over €2,000 billion by 2010, including €820 billion in the pharmaceutical sector (EC, 2002). Not surprisingly, many national governments are eager to gain a stake in this market and reap the fruits of academic research and have, therefore, been supporting the establishment of new companies. In the late 1990s, the Dutch government introduced the Action Plan Life Sciences, having realized that—despite the country's abundance in new knowledge production and well-developed infrastructure for clinical trials—new entrepreneurship in the area of biotechnology was lagging behind other countries (MEA, 2000). Furthermore, the government recognized that entrepreneurial culture in the Netherlands in general was only mediocre due to unnecessary regulatory constraints, such as high costs for establishing a limited liability company and for hiring and firing employees. To stimulate entrepreneurial culture in the Netherlands, the government introduced measures to simplify the rules and laws for establishing a new company, reduced the costs involved, etc. (MEA and EIM, 2001).

This chapter examines to what extent the Action Plan Life Sciences has promoted institutional transformation in the Netherlands and whether it fostered the growth of biotechnology companies. The policy was successful in principle, but produced a large number of relatively vulnerable, small firms. The chapter begins with the 1998 findings about institutional shortcomings in the Netherlands and the policy implemented to respond to these from 2000 onwards. The policy implemented by the government places an implicit emphasis on university-based clusters, which is also characteristic of many other national policies (Zeller, 2001). Subsequently, the structure of the biotechnology "landscape" in the Netherlands is examined to determine whether the adoption of a cluster-oriented policy is useful. This is followed by a critical evaluation of the policy since its implementation, based on perceivable changes within the sector and in the relevant institutions. The chapter concludes with a summary of the results and some recommendations for future research.

2 Action Plan Life Sciences

The results of an international benchmark study comparing the Netherlands with six international regions—the Boston area (United States), Cambridge (United Kingdom), Quebec (Canada), Munich (Germany), Flanders (Belgium) and Skiruna (Denmark/Southern Sweden)—revealed several institutional shortcomings in the Netherlands, including most importantly its weak entrepreneurial culture and low quality of incubators (MEA, 1999; van Geenhuizen, 2003). Weaknesses were also identified in the entrepreneurial culture of academic researchers, and particularly, in the risk-aversive policies of universities. Other major deficiencies included the low quality of initiatives and plans for the establishment of new companies, attributable to the exclusively technical backgrounds of company founders; the insufficient quality of incubator sites connected to universities including accommodation and supporting infrastructure; the low availability of pre-seed and seed capital in a country in which venture capital providers tend to only invest in later stages; and an inadequate matching of science and business as a result of a lack of networks in which key actors can collaborate.

The Action Plan Life Sciences was initiated in response to these inefficiencies. The policy reflects the principle of differentiating financial support in accordance with the company's development phase and early growth of start-ups. Hence, the Action Plan's comprehensive approach builds on five distinct programs (Table 1). More precisely, the Plan's objective was to support the establishment of 75 biotechnology companies from 2000 to 2004, with a total budget of €45 million. Though €45 million may at first glance be considered a modest sum, the Dutch economy's relatively small size with a GDP of $434 billion in 2003 must be taken into consideration, as well as the country's size, which is comparable with some of the largest biotechnology regions in the world, like the Boston area and the San Francisco Bay area in the USA.

An intermediary organization, BioPartner, was set up to facilitate the implementation of the Action Plan's five programs. BioPartner furthermore functioned as a learning organization with the task of surveying the market and the technological

Table 1. The Action Plan Life Sciences Programs

Programs and Available Budget	Purpose
First Stage Grant (pre-seed fund) *€11* million	To encourage researchers to carry out applied research and apply for patents. It covers researchers' salaries, costs of patenting, etc.; up to €250,000 for a maximum of 2.5 years.
Facilities Support (facility fund) *€5* million	To provide loans to universities and research organizations to pre-finance advanced life sciences facilities and to provide start-ups with the possibility of acquiring advanced equipment at a reasonable price. Later, a facility leasing program was launched as an alternative.
Start-up ventures (seed fund) *€11* million	To provide investment capital on the condition that private parties also make a substantial (equal) investment. The fund invests up to a maximum of €227,000 in start-ups, and an additional €227,000 may be made available in a subsequent round of financing.
BioPartner Centers (six incubators) *€11* million	To provide laboratory, office, R&D, pilot plant facilities, and shared services for university-linked incubators, and to facilitate access to universities and other companies (in Leiden, Amsterdam, Groningen, Utrecht, Wageningen, and Maastricht).
BioPartner Network *€7* million	To facilitate and stimulate entrepreneurship through advice, information, scouting, training programs, and creating networks, e.g., the establishment of special chairs and introduction of additional postgraduate courses at universities.

Source: adapted from MEA (2000) and BioPartner (2005)

environment of biotechnology; monitoring the programs' effectiveness and, if necessary, modifying the program's practical application; and monitoring new entrants in terms of their structural characteristics, product pipeline, financial expectations, etc. The Action Plan Life Sciences had an implicit focus on spatial clusters around universities, as evidenced by the locations selected for the six BioPartner incubators, which are in close proximity to universities and their facilities (in most cases, research medical centers). The next section focuses on the spatial pattern of biotechnology in the Netherlands.

3 The Biotechnology 'Landscape' in the Netherlands

Research-intensive biotechnology companies are—more so than other companies in the high technology industry—clustered around public scientific organizations, including universities, research medical centers, and laboratories that provide knowledge and facilities, or (predominantly found in the USA) venture financing (Audretsch, 2001; Cooke, 2001, 2004; Powell et al., 2002). The notion that a clustered location provides advantages is widely accepted; this holds particularly true for benefits gained from local knowledge networks and knowledge spillovers from local universities and research medical centers, although recent research findings question whether local knowledge networks are in fact the dominant contributors to growth and innovativeness (Gertler and Levitte, 2005; van Geenhuizen, 2007).

The largest single biotechnology cluster in the Netherlands is located in the Leiden region in the Western part of the country between Amsterdam and The Hague, which, like Leiden, belong to the Randstad conurbation. We did not use official industrial statistics to measure the size of individual clusters, because the companies included in such statistics are scattered across different sectors. Therefore, we used reports on the life sciences sector, which generally cover the majority of companies belonging to such clusters (BioPartner, 2001, 2002, 2003, 2004, 2005). With 25 dedicated life sciences firms in 2005,[2] Leiden's cluster is larger than those in other cities in the Netherlands, like Amsterdam and Groningen, whose clusters consist of 15 to 20 companies. The Leiden cluster comprises approximately 40 companies in total, when including foreign subsidiaries, pharmaceutical companies, medical equipment companies, and consultancies. The origin of Leiden's biotechnology cluster goes back to the early 1980s, and gained momentum when the municipality established a Science Park in 1984 in collaboration with the university, which included an incubator facility. Today, the cluster consists of a number of faculties of the University of Leiden (including its medical school and research medical center), two higher education institutions, the Netherlands Organization of Applied Research (TNO) in the field of health care, and two national research centers, one of which is specialized in genomics. From an international perspective, the size of the Leiden cluster may seem relatively small, but should be considered in the context of Leiden's location in the Randstad conurbation, which is one of the largest in Europe. The biotechnology clusters of Amsterdam (30 km), Utrecht (40 km), Delft (35 km), and Rotterdam (45 km) are within close proximity of Leiden.

Thus, if the group of clusters within the Randstad conurbation is taken as one, the number of dedicated biotechnology firms amounts to 70, which is still relatively small compared to biotech mega centers around the globe (Cooke, 2004).

This section examines whether a clustered location contributes to the innovation level of companies in the vicinity in order to identify arguments in favor of a cluster policy. We draw on an estimation of a regression model of innovativeness, including various company characteristics, i.e., age, size, mode of founding, and location as independent variables (van Geenhuizen and Reyes-Gonzalez, 2007) (Table 3). The model was limited to one available input indicator of innovativeness, namely R&D employment.[3] We used a narrow classification of innovativeness, since output

[2] Dedicated life sciences firms are newly established, entrepreneurial companies with a focus on the life sciences; this definition excludes diversified companies, foreign subsidiaries, consultancies, and medical equipment companies.

[3] In most studies, R&D employment is taken as a percentage share of total employment. We have, however, experimented using both absolute numbers, as well as percentage shares. The disadvantage of using percentage shares in our study is that many small companies have an extremely high share of employees engaged in R&D, because they often focus exclusively on R&D, particularly at the onset, while larger companies usually have a relatively low share of employees in R&D, but have a higher number of such employees. We therefore used absolute numbers as well. In addition, we concluded that the critical mass in companies' in-house R&D ought to be depicted in absolute size (and not in shares).

Table 2. Biotechnology in the Netherlands (2005)[a]

Location	No. of Companies	Description
Leiden (largest single cluster)	25	Within a range of approximately 5 km from knowledge institutes in Leiden. A strong focus on general and medical biotechnology.
A group of clusters (Randstad), Leiden, Amsterdam, Utrecht, Delft, and Rotterdam	70	The maximum distance between the group of clusters is approximately 65 km (from Amsterdam to Rotterdam). A strong focus on general and medical biotechnology.
The Netherlands	165	The maximum distance between the national clusters is about 300 km (from Groningen to Maastricht). A focus on general and medical biotechnology, as well as agro-food biotechnology (mainly in Wageningen).

[a] Dedicated life sciences firms
Source: Adapted from BioPartner (2005)

indicators of the innovation process, like patents and new products introduced to the market, fall beyond the scope of this study. The data used refer to the situation in 2002/3, i.e., approximately midway through the Action Plan's time-span. The number of dedicated life sciences firms at the time was 126, of which 96 were included in the regression analysis on the basis of full data availability. "Location" in Model 1 distinguishes between a clustered location versus a non-clustered one, whereas in Model 2, 'location' refers to a clustered location in the Leiden area versus all other areas. A positive sign of the coefficient was expected for age and size, and a negative one for independently established companies, because of the assumed lack of support from a parent organization. A positive sign was also expected for a clustered location in both models.

However, the estimation results with regard to age did not correspond with our assumption, because not only did the sign differ, but the coefficient was significant as well. Apparently, biotechnology companies start as highly innovative ventures engaged in R&D, and as they mature, their R&D activity stagnates. Such a development is most likely caused by the lack of investment capital, a common development following the initial growth stage (Ernst & Young, 2005). An alternative and less pessimistic explanation is that companies with insufficient capital at their exposure stabilize or reduce their R&D activity and establish collaborative alliances with larger companies or research institutes, of which many are located abroad. Additional research on reasons why biotechnology companies establish partnerships with other companies and research institutes also confirms this trend (BioPartner, 2005). The estimation results for size corroborated our expectations of a positive sign and a high significance, indicating that the larger a company is, the more R&D is carried out. This impact of size on R&D (measured by absolute numbers of jobs) is not self-explanatory. It seems that at the onset, biotechnology companies conduct

R&D activities only, and that—as they mature—R&D jobs grow at the same rate. This implies that the larger entrepreneurial companies have not yet reached the stage of maturity and are concentrating on production, but actually remain engaged in R&D.

Table 3. Regression Results

	Model 1		Model 2	
	Coefficient	P-value	Coefficient	P-value
Constant	4.32	0.053	3.90	0.066
Age	–0.32	0.027 *	–0.32	0.022 *
Size	0.70	0.000 **	0.70	0.000 **
Corporate Spin-off	3.76	0.090	–3.34	0.121
Academic Spin-off	–5.01	0.013 *	–4.17	0.036 *
IEV[a]	–3.21	0.105	–2.84	0.139
Clustered	0.86	0.386	n.a.	
Clustered in Leiden	n.a.		2.77	0.020 *
R^2	0.957		0.956	

[a]Independently Established Ventures
significant at 0.95 level; **significant at 0.99 level.
N (companies): 96
Source: Adapted from van Geenhuizen and Reyes-Gonzalez (2007)

With regard to corporate position, the coefficient was only significant for academic spin-offs. The negative sign of the coefficient implies that academic spin-offs tend to have fewer jobs in R&D, a circumstance that may be explained by the relationship between academic spin-offs and universities, where part of the research is carried out. Finally, the coefficient for a clustered location in Model 1 was positive, but unlike our initial expectation, it was insignificant. This indicates that a clustered location is not relevant for R&D. In Model 2, however, the coefficient for a clustered location in the Leiden area revealed not only a positive sign, but turned out to be significant as well, implying that a clustered location in Leiden tends to contribute to enhanced innovation performance, in comparison to companies in other locations. Apparently, in terms of location, the largest cluster seems to provide a number of *unique* and valuable qualities.

To identify these qualities, mixed evidence was collected, i.e., in-depth interviews with corporate managers and experts in this sector were carried out (van Geenhuizen and Reyes-Gonzalez, 2007). One of the advantages of the Leiden area is the easy access to the international airport of Amsterdam Schiphol (a 15–20 minute drive), which plays an important role for companies interested in collaborating with international partners. This advantageous location is, however, applicable to Amsterdam and Utrecht as well. Another benefit is the Leiden area's sheer size and range of possibilities. Services from specialized companies are available, as is the prospect of pooling resources with different laboratories. Leiden's labor market is also better developed than that of other regions. Owing to its long experience as a cluster, the Leiden region enjoys a strong local accumulation of experience-based knowledge with regard to start-up strategies and global partnerships. In terms of knowledge networking, local networks only play a significant role during the inception and incubation stage, but not in the later growth stages of research start-ups (van Geenhuizen, 2007). Unfortunately, no similar studies in other European countries have been carried out to date, which would provide an opportunity to compare our results.[4]

4 Biotechnology Developments from a Policy Perspective

This section examines the developments in the biotechnology sector of the Netherlands following the implementation of the Action Plan Life Sciences. Most data used represent the life sciences on the whole, which is why this term is used here, as well as in the subsequent sections. It must be noted that a methodologically sound evaluation in the sense of establishing causal relations between the policy and actual developments was not possible. The main reasons are that we cannot determine how the life sciences industry would have developed without the launch of the Action Plan, and that some years in which impacts of the policy may still have unfolded, fall beyond the time period covered (from 2000 to 2004) by our study. Given this limitation, the focus of attention was the growth of the sector and its structural change.

A clear increase in the number of new start-ups in the life sciences industry throughout the policy program's implementation is evident. From 2000 to 2004, the index number of new start-ups increased to 244.4, bringing the total number of dedicated life sciences firms to nearly 160 (Table 4). The launch of the program in 2000 anticipated the establishment of 75 new life sciences companies within the Plan's five-year time frame, yet in fact, initial expectations were surpassed with a total of 102 new companies being set up. Considering that the program was largely implemented during years of general macro-economic downturn—which is reflected in the decrease in the overall number of new start-ups below the level of 1998—the quantitative results of life sciences companies are even more impressive. In addition, the survival rate among these entrants was quite high (84%), obviously due to the protective environment provided by the program in the start-ups' early growth stage (BioPartner, 2004, 2005).

In a preliminary attempt of curve fitting (on the basis of seven years), the linear and polynomial models tend to be the most suitable growth models for new start-ups in the life sciences (Table 5). The polynomial model reveals a slight decrease in growth during the years of economic recession (2001–2003), followed by a renewed increase. This growth pattern confirms the vulnerability of the life sciences sector to the ups and downs of the macro-economy. The question arises whether this growth

[4] A study on Canada has been carried out, but is problematic in terms of spatial delineation of clusters (Niosi and Banik, 2005).

Table 4. Total Number of Dedicated Firms and New Start-Ups in the Life Sciences Industry and the Economy at Large

Year	Total number of life sciences firms (index numbers in brackets)	New start-ups in the life sciences per year (index numbers in brackets)	New start-ups in all sectors per year (index numbers in brackets)
1998	57 (100.0)	9 (100.0)	71.463 (100.0)
1999	74 (129.8)	14 (155.6)	77.797 (108.9)
2000	97 (170.2)	24 (266.7)	88.014 (123.2)
2001	118 (207.0)	21 (233.3)	80.269 (112.3)
2002	126 (221.1)	18 (200.0)	68.961 (96.5)
2003	138 (242.1)	17 (188.9)	68.947 (96.5)
2004	157 (275.4)	22 (244.4)	76.252 (106.7)

Source: BioPartner (2001, 2002, 2003, 2004, 2005); MEA (2003a,b); EIM (2006)

Table 5. Growth Models of New Start-Ups in the Life Sciences (1998–2004)

Linear model	Polynomial model
$Y = 16{,}321X + 44{,}286$	$Y = -0{,}8452X^2 + 23{,}083X + 34{,}143$
$R^2 = 0.9833$	$R^2 = 0.9912$

pattern would have been more pronounced had the Action Plan Life Sciences not been introduced.

It must also be noted that the increase in the number of start-ups in the life sciences industry was not coupled with a similar increase in employment (112%) and turnover (164%) (Table 6). In other words, the sector owed its growth primarily to the small business segment, a development confirmed by the data on size structure. The majority of dedicated life sciences firms employed less than ten full-time employees (FTE), a share that increased between 2000 and 2004 from 65% to 73% (Table 7).

This, linked with the sector's net result of a loss of €75 million in 2004, clearly indicates that the life sciences industry is at an early stage of development and is dominated by highly vulnerable companies. In addition, a shift in business models has taken place toward product companies (from 11% to 27%), resulting in a higher level of innovativeness (and risk), and, at the same time, toward hybrid companies (from 49% to 57%). In fact, the hybrid model has been adopted by the majority of companies (Table 7). This shift toward a larger segment of hybrid companies is reflected in the relatively small increase of R&D expenditure in the years 2000–2004, compared with the rise in the number of companies (115% versus 175%).

The distinctive development of the Netherlands' life sciences industry is also visible in comparison with other European countries (Table 8). In terms of number of companies in 2004, the Netherlands ranked among the top five, but average employment within these companies was the lowest among the top ten, and half the size of the average for all of Europe (22.9 versus 44.6). This stands in contrast with the results of other small economies, like Switzerland and Denmark, which can be

Table 6. Growth Indicators of Dedicated Life Sciences Firms

Type of Result	Period/Year	Outcome
Companies and employment		
Total number of companies	2004	157
Total increase in companies	2000–2004	+175.4%
Number of new start-ups	2000–2004	109 (22 on average)[a]
Total increase in start-ups	2000–2004	+319%
Number of start-up survivors	End of 2004	92 (84%)
Employment in sector	End of 2004	2.150 FTEs
Total increase in employment	2000–2004	+112%
Financial		
Turnover	2004	190 million
Total increase in turnover	2001–2004	+164%
Net result	2004	–75 million
R&D expenditure	2004	118 million
Total increase in R&D expenditure	2000–2004	+115%

[a]Due to definitional differences, the total over five years is slightly higher than the number of new start-ups in each separate year, as recorded in annual reports (see Table 4).
Source: Adapted from BioPartner (2005)

Table 7. Size and Business Models among Dedicated Life Sciences Firms

Type of Result	Period/Year	Outcome
Size structure (% share)		
<10 FTE		73%
11–25 FTE	End of 2004	14%
26–50 FTE		6%
>50 FTE		7%
Companies < 10 FTE (change)	2000–2004	from 65% to 73%
Business model (% share)		
Service company		8%
Tool company	End of 2004	8%
Product company		27%
Hybrid company		57%
Service company (change)	2000–2004	from 33% to 8%
Product company (change)	2000–2004	from 11% to 27%
Hybrid company (change)	2000–2004	from 49% to 57%

Source: Adapted from BioPartner (2005)

explained by the Netherlands' lack of life sciences divisions in large pharmaceutical companies, as well as the country's exceptionally high number of very small dedicated life sciences firms.

A closer look at some of the characteristics of young innovative companies in the Netherlands, mainly active in the medical sector, reveals that only a few are public enterprises and that most of them are in an early stage of the product-pipeline (preclinical testing) (Table 9). Venture capital investment is also relatively low in the Netherlands. This further corroborates the fact that the Netherlands' life sciences

Table 8. Biotechnology in Europe (Ten Countries with the Largest Biotech Sectors) (2004)[a]

Country	Companies	Employment	Average Employment per Company	Young Innovative Companies <5 Years (%)
Germany	538	16,094	29.9	61
United Kingdom	457	21,134	46.2	55
France	233	9,142	39.2	60
Sweden	138	3,942	28.6	61
Netherlands	124	2,837	22.9	64
Denmark	117	18,461	157.8	68
Switzerland	90	4,990	55.4	70
Belgium	84	3,654	43.5	51
Spain	81	2,201	27.2	60
Finland	66	2,160	32.7	45
Europe	2,163	96,459	44.6	59

[a]The definition of biotechnology in this table is slightly different from that used in studies by the Dutch BioPartner, as established biotechnology companies are included here as well.
Source: Adapted from Critical 1/EuropaBio (2006)

Table 9. Public Enterprises' Product Pipeline (Ten Countries with the Highest Product Pipeline) and Venture Capital Investment in the Sector (2005)

Country	Product Pipeline Pre-Clinical	Product Pipeline Clinical (I–III)	Venture Capital Investment (% of GDP)
United Kingdom	64	147	0.016
Switzerland	51	58	0.009
Denmark	19	31	0.024
Germany	15	24	0.005
France	14	22	0.007
Sweden	8	17	0.015
Austria	6	4	0.005
Belgium	5	3	0.005
Finland	5	2	0.009
Netherlands	4	2	0.003

Source: Adapted from Ernst & Young (2006) (product pipeline) and OECD 2006 (venture capital)

industry is made up of young and vulnerable companies. A comparison with other small economies, for example, Switzerland, Denmark, and Sweden, is noteworthy. The presence of a strong domestic pharmaceutical (or biochemical) industry, actively engaged in innovation, seems to make all the difference. Such an industry has largely been absent in the Netherlands, or has moved parts of its R&D abroad. However, DSM and AKZO-Pharma, two medium-sized Dutch enterprises, have recently assumed a supportive role to assist small companies in reaching the next stages of product development. Likewise, the Dutch life sciences firms, Crucell and Pharming, also offer support to small companies.

When looking at the recent past, we can conclude that the life sciences industry in the Netherlands is in an early growth stage. Thus, the sector today primarily consists of small, loss-making, entrepreneurial companies with fewer than ten employees. The number of such companies rapidly increased during the Action Plan's implementation period, essentially enlarging the segment of very small firms and leading to a segment dominated by hybrid companies. This outcome raises some interesting policy questions, particularly about the transformation of institutions and about institutions that did not change. These are addressed in the following section.

5 Institutional Transformation?

When considering the Action Plan Life Sciences from the perspective of institutional transformation, the following four observations can be made (van Geenhuizen, 2009):

1. A successful transformation of institutions took place, which has had a positive effect on academic entrepreneurship and on the granting of seed-money.
2. The transformation of institutions did not have an effect on the supply of venture capital; only some progress was evident.
3. The policy of spreading incubators across the country reflects the principle of equity, an outdated principle that has already been abandoned in current regional-economic policy (MEA, 2004).
4. Limitations for entrepreneurship attributable to other policy areas remained beyond the Action Plan's scope (mainly restrictive regulation).

These four issues will now be discussed in more detail. The Action Plan Life Sciences was able to contribute to the reduction of institutional obstacles in academic entrepreneurship. In terms of entrepreneurial culture, awareness has been increased and universities have adopted a more pro-active attitude with regard to the commercialization of knowledge. In terms of financial support, success was documented in the first stage (seed-stage), in which academics translate their knowledge into a feasible business plan. 128 applications for seed-funds were submitted in the past four years (2001–2004) and 67 (52.3%) grants were awarded. A requirement with regard to the financing of ventures, however, was that private funds, equaling the amount granted by the Action Plan's program, had to be found. This policy program turned out to be less successful compared with the seed funding program. 224 applications were submitted of which 68 (30.4%) were approved, yet only 38 (17.0%) actually materialized. However, because of the time needed by start-ups to find private funds equaling a specific amount, a longer period of observation than was included here would be required to determine the total number of approvals.

We can conclude that the small number of cofinanced ventures to date indicates an ongoing reluctance by the venture capital market to invest in relatively early and risky stages of life sciences ventures (equity gap).[5] The Action Plan Life Sciences did not improve this situation. The venture capital sector in general has been more reluctant and has refrained from high-risk investments due to the economic downturn in the recent past. Today, however, some progress is unfolding. It seems that the widespread equity gap problem can be sidestepped in selected start-ups, which are provided with capital by venture capitalists at an early stage, who, however, also withdraw from the venture relatively quickly.[6]

The third issue addressed concerns the spread of the six incubators across the country. This pattern reflects an institutional trend in the Netherlands' regional economic policy of the 1980s in which the guiding principle was equity, and not efficiency in terms of intensifying developments that are already strong, as is the case today. As a result, the Action Plan's policy may have missed the chance to generate the critical mass needed for a strongly competitive market, and may have even prompted duplication among different clusters. If, for example, only three incubators had been set up, each with a strong focus on technology, critical mass may have been generated more quickly and the incubators' competitive power and international image may have become much stronger. The fact that the Leiden cluster, which offers more and better opportunities than other clusters, was not prioritized in the Action Plan corresponds with the equity trend of the 1980s. In other words, policymakers resorted to an outdated institutional trend. Furthermore, given the fact that six regional centers were selected, cooperation between them could have been stimulated from the start to develop a strong, interactive network of clusters based on complementarity. In addition, whether the model of supplying accommodation (e.g., the BioPartner incubators), without having precise information on actual demand, is efficient must be questioned as well. There are indications that some incubators are having difficulty finding biotechnology start-ups as tenants (MEA, 2005).

The Action Plan Life Sciences has contributed to significant growth in the life sciences sector, mainly through an increase in the segment of small and vulnerable companies. It is not clear whether the Netherlands will be able to seize the benefits of recent developments and whether the sector is able to progress to a mature stage and promote the establishment of several medium-sized and large life sciences companies. The focus of attention in the remaining section will consequently be on the need for a supportive policy for later development stages, with the objective of preventing a shakeout among small and vulnerable companies or encouraging small companies to grow, as well as endorsing their R&D efforts. Support for smaller firms' R&D activities is a realistic option, as suggested by the results of our regression analysis on R&D.

The final issue concerns regulatory conditions that may hamper entrepreneurship, but fall beyond the scope of the Action Plan's policy program. It seems that regulations on genetic engineering and field testing in agricultural production, as well as on the breeding of genetically modified animals to produce pharmaceutical

---

[5] The equity gap refers to the fact that the period of financing necessary for the commercialization of new medicines is longer than the years for which venture capitalists would typically provide money.

[6] Source: private communication.

substances, are more stringent and rigid in the Netherlands than in other European countries. Several companies have shut down operations in the Netherlands or relocated part of their activities to other countries, given that the implementation of field tests may invoke negative reactions among certain segments of the public (including the destruction of test fields) (MEA, 2005). This constraint was not explicitly addressed in the Action Plan, but will be addressed in a follow-up policy.

6 Supportive Policies for Later Growth Stages

An essential survival strategy of biotechnology start-ups is an early R&D partnership with other companies or research institutes. Partnerships with larger companies, however, are not necessarily a bed of roses. It is difficult to select the most suitable partner, to identify the best point in time for collaboration given the different test and trial stages, and to negotiate the most appropriate terms for collaboration, but also for ending the partnership once the goals have been achieved. Because the Netherlands does not have a large, domestic pharmaceutical industry and its biotechnology companies are still comparatively weak, global partnerships—at least in the medical/pharmaceutical segment—are frequently set up (van Geenhuizen and Knaap, 1997; van Geenhuizen and Reyes-Gonzalez, 2007). As indicated in the previous section, collaboration between domestic partners has recently been increasing.

National programs can create a supportive environment and encourage partnerships by coaching small ventures in the establishment of global partnerships and, secondly, by creating better opportunities for early-stage research partnerships at home. This second recommendation has partially been embraced by the recently established Top Institute Pharma (TI Pharma) in Leiden. It is, however, limited to bio-pharmaceutical companies. Although the financial conditions for partnership are, strictly speaking, too stringent for small companies, they have the possibility of participating in the Institute under less stringent financial regulations. TI Pharma's mission is to take over the leadership in research areas critical for the Dutch pharmaceutical industry's international competitiveness, by connecting and merging results from presently fragmented disciplines, programs, and locations. TI Pharma's institutional model builds on the Technological Top Institute, an internationally acclaimed "best practice" model of public-private cooperation that complements the increased relevance of networking, demand-driven research, and program financing. TI Pharma is a consortium of participating companies (a mix of large and medium-sized ones, foreign subsidiaries and domestic, independent companies), universities, research institutes, and three Ministries (Health, Education, and Economic Affairs), all of which contribute to TI Pharma's budget. The features that were missing in the Action Plan—the coordination of different clusters, the creation of a critical mass, and new modes of research financing—are now available in the pharmaceutical industry within a new institutional model.

The Action Plan's program only lasted five years. In its self-evaluation study, the Ministry of Economic Affairs (MEA, 2005) concluded that the five year period was too short to achieve any long-term results, particularly because far-reaching institutional transformation requires far more time, as does the commercialization process (e.g., in the case of new drugs). At present, the government is reflecting on a follow-up policy, which will include both specific (life sciences) and generic elements (all technology sectors). This policy would, in particular, focus on ways to prevent or bridge the equity gap (van Geenhuizen, 2009).

7 Discussion

This chapter on life-sciences policy used the Netherlands as an example of a small economy, characterized by a risk-averse culture at universities, and a lack of venture capital in the early growth stages of start-up companies in the late 1990s. The Netherlands provides an interesting case study for policymakers, because the country lagged behind other countries in the commercialization of knowledge, despite the strong scientific base of its universities. The country also lacked a strong, domestic pharmaceutical industry in innovative segments capable of propelling small start-ups to the subsequent stages of knowledge commercialization. The national policy approach was quite comprehensive in its response to this situation, which included the improvement of the sector's image and easing of requirements for receiving seed-money, as well as establishing and connecting crucial networks to foster the collaboration of key actors. In terms of number of life sciences companies, improvements are clearly visible—the Netherlands is ranked among the top five in Europe. But in terms of young, innovative companies' competitive strength, the situation continues to pose a challenge. Furthermore, the strategy of spreading new incubators to six different regions across the country is ambiguous, particularly because innovation tends to be especially strong in the largest cluster. Overall, the policy appears to have been successful in transforming institutions to foster academic entrepreneurship, but failed to a certain extent to transform institutions to modify the supply of venture capital and to support smaller firms in their quest to enter subsequent growth stages. Although some institutional adjustments are currently under way, taking account specifically of the fairly new model of the Technological Top Institute, it remains unclear whether the segment of small life sciences companies can actually progress to a mature stage in the near future.

Close monitoring of small companies will continue to remain vital in upcoming years. Research on how such companies overcome hurdles on their way to the next growth stage(s) and what the reasons may be for their failure to reach the following stage could provide valuable insights. A comparative study of the Netherlands with countries that have successfully eliminated institutional barriers to the funding of early growth and later growth stages of small life sciences companies would also be particularly useful in this context.

References

Audretsch D (2001) The role of small firms in U.S. biotechnology clusters. Small Business Economics 17:3–15

Audretsch D B, Keilbach M C, Lehman E E (2006) Entrepreneurship and economic growth. Oxford University Press, Oxford

BioPartner (2001) The emerging Dutch life science sector. Biopartner, Ede

BioPartner (2002) The Netherlands life sciences sector report 2002—Gaining momentum. BioPartner, Ede

BioPartner (2003) The Netherlands life sciences sector report 2003—Growth against the tide. BioPartner, Ede

BioPartner (2004) The Netherlands life sciences sector report 2004—Moving forward. BioPartner, Ede

BioPartner (2005) The Netherlands life sciences sector report 2005—New challenges ahead. BioPartner, Ede

Cooke P (2001) Biotechnology clusters in the UK: Lessons from localisation in the commercialisation of Science. Small Business Economics 17:43–59

Cooke P (2004) The molecular biology revolution and the rise of bioscience megacentres in North America and Europe. Environment and Planning C: Government and Policy 22:161–177

Critical 1/EuropaBio (2006) Biotechnology in Europe. 2006 comparative study. Critical 1/EuropaBio, Banbury and Brussels

EIM (Economic Institute of Small- and Medium-Sized Enterprises) (2006) International benchmark 2005; Benchmark on entrepreneurship, firm dynamics and quickly growing companies. EIM, Zoetermeer

Ernst & Young (2005) Beyond borders. Global biotechnology report 2005. Ernst & Young

Ernst & Young (2006) Beyond borders. Global biotechnology report 2006. Ernst & Young

EC (European Commission) (2002) Life sciences and biotechnology. A strategy for Europe. COM (2002) 27. European Communities, Luxembourg

Geenhuizen M van (2003) How can we reap the fruits of academic research in biotechnology? In search of critical success factors in policies for new-firm formation. Environment and Planning C 21:139–155

Geenhuizen M van (2007) Modelling dynamics of knowledge networks and local connectedness: A case study of high-tech companies in the Netherlands. Annals of Regional Science 41:813–833

Geenhuizen M van (2009) Life sciences in the Netherlands: Weakness in a strong policy. European Planning Studies (forthcoming)

Geenhuizen M van, Knaap B van der (1997) Dynamics of R&D and regional networks in the Dutch pharmaceutical industry. Journal of Social and Economic Geography 88:307–320

Geenhuizen M van, Reyes-Gonzalez L (2007) Do clusters improve performance in biotechnology? A look at the biotechnology industry in the Netherlands. Technological Forecasting and Social Change 74(9):1681–1696

Gertler M, Levitte Y M (2005) Local nodes in global networks: The geography of knowledge flows in biotechnology innovation. Industrial Innovation 12(4):487–507

Malerba F (2006) Innovation and the evolution of industries. Journal of Evolutionary Economics 16:3–23

McKelvey M, Orsenigo L, Pammolli F (2004) Pharmaceuticals analysed through the lens of a sectoral innovation system. In: Malerba F (ed) Sectoral systems of innovation, Cambridge University Press, Cambridge

MEA (Ministry of Economic Affairs) (1999) Letter of the Minister of Economic Affairs to the parliament concerning an action plan for new firms in the life sciences (in Dutch). Ministry of Economic Affairs, The Hague

MEA (Ministry of Economic Affairs) (2000) Action Plan Life Sciences 2000–2004. Ministry of Economic Affairs, The Hague

MEA (Ministry of Economic Affairs) and EIM (Economic Institute of Small- and Medium-Sized Enterprises) (2001) Entrepreneurship in the Netherlands. New economy: New entrepreneurs! MEA-EIM, The Hague-Zoetermeer

MEA (Ministry of Economic Affairs) (2003a) Entrepreneurship in the Netherlands. Knowledge transfer: Developing high-tech ventures. Ministry of Economic Affairs, The Hague

MEA (Ministry of Economic Affairs) (2003b) Life sciences. A pillar of the Dutch knowledge economy. Ministry of Economic Affairs, The Hague

MEA (Ministry of Economic Affairs) (2004) Peaks in the Delta. Ministry of Economic Affairs, The Hague

MEA (Ministry of Economic Affairs) (2005) Life science monitor 2005. Status of the Life Science Action Plan of the Ministry of Economic Affairs. Ministry of Economic Affairs, The Hague

Niosi J, Banik M (2005) The evolution and performance of biotechnology regional systems of innovation. Cambridge Journal of Economics 29:343–357

OECD (2006) OECD biotechnology statistics. OECD Publishing, Paris

Powell W W (1998) Learning from collaboration. Knowledge and networks in the biotechnology and pharmaceutical industries. California Management Review 40(3):228–240

Powell W W, Koput K W, Bowie J I, Smith-Doehrr L (2002) The spatial clustering of science and capital: Accounting for the biotech firm-venture capital relationships. Regional Studies 36:291–305

Salman N, Saives A-L (2005) Indirect networks: An intangible resource for biotechnology innovation. R&D Management 35(20):203–215

Senker J, Sharp M (1997) Organisational learning in cooperative alliances: Some case studies in biotechnology. Technology Analysis and Strategic Management 1:35–51

Walsh V (1993) Stimulating demand. Demand, public markets and innovation in biotechnology. Science and Policy June 1993:138–156

Zeller C (2001) Clustering biotech: A recipe for success? Spatial patterns of growth of biotechnology in Munich, Rhineland and Hamburg. Small Business Economics 17:123–141

Index

Note: Page numbers in bold refer to Tables; those in italics refer to Figures

The International Institute for Applied Systems Analysis

is an interdisciplinary, nongovernmental research institution founded in 1972 by leading scientific organizations in 12 countries. Situated near Vienna, in the center of Europe, IIASA has been producing valuable scientific research on economic, technological, and environmental issues for over three decades.

IIASA was one of the first international institutes to systematically study global issues of environment, technology, and development. According to IIASA's Governing Council, the Institute's goal is: *to conduct international and interdisciplinary scientific studies to provide timely and relevant information and options, addressing critical issues of global environmental, economic, and social change, for the benefit of the public, the scientific community, and national and international institutions.* Research is organized around three central themes:

- Energy and Technology
- Environment and Natural Resources
- Population and Society

IIASA is funded and supported by scientific institutions and organizations in the following countries: Austria, China, Egypt, Estonia, Finland, Germany, India, Japan, Republic of Korea, Netherlands, Norway, Pakistan, Poland, Russian Federation, South Africa, Sweden, Ukraine, United States of America.

Further information: http://www.iiasa.ac.at

